HISTORY OF THE SCIENCES IN
GRECO-ROMAN ANTIQUITY

BIBLO & TANNEN PUBLICATIONS

Abbott, F. F.:   Roman Political Institutions
Abbott, F. F.:   Society & Politics in Ancient Rome
Budge, E. A. W.:   The Mummy
Cave, R. C.:   Medieval Economic History
Childe, V. G.:   The Bronze Age
Clark, E. D.:   Roman Private Law — 4 vols.
Cook, A. B.:   Zeus. Vol. 1
Cook, A. B.:   Zeus. Vol. II — 2 vols.
Davis, W. S.:   A Day in Old Athens
Davis, W. S.:   A Day in Old Rome
Dodge, T. A.:   Caesar — 2 vols.
Evans, A.:   Palace of Minos at Knossos — 4 vols. in 7
Ferguson, W. S.:   Greek Imperialism
Hamburger, M.:   Morals and Law
Hasebroek, J.:   Trade & Politics in Ancient Greece
Henderson, E. F.:   Select Historical Documents of the Middle Ages
Judson, H. P.:   Caesar's Army
Minns, E.:   Scythians and Greeks — 2 vols.
Pausanias   Description of Greece — Trans. J. G. Frazer 6 vols.
Pendlebury, J. D. S.:   The Archaeology of Crete
Petersson, T.:   Cicero. A Biography
Power, E.:   English Medieval Nunneries
Powicke, F. M.:   Ways of Medieval Life & Thought
Previte-Orton, C. W.:   Medieval History
Reymond, A.:   Hist. of Sciences in Greco-Roman Antiquity
Rostovtzeff, M. I.:   Out of the Past of Greece and Rome
Sayce, R. U.:   Primitive Arts and Crafts
Scott, J. A.:   Unity of Homer
Seymour, T. D.:   Life in the Homeric Age
Smyth, H. W.:   Greek Melic Poets
Thomson, J. O.:   History of Ancient Geography
Tozer, H. F.:   A History of Ancient Geography
Wace, A. J. B.:   Mycenae

HISTORY OF THE SCIENCES IN GRECO-ROMAN ANTIQUITY

BY

ARNOLD REYMOND

PROFESSOR OF PHILOSOPHY AT THE UNIVERSITY OF LAUSANNE

TRANSLATED BY

RUTH GHEURY DE BRAY

WITH 40 DIAGRAMS

BIBLO and TANNEN
New York
1965

BIBLO and TANNEN
BOOKSELLERS and PUBLISHERS, Inc.
63 Fourth Avenue   New York 3, N.Y.

Library of Congress Catalog Card Number: 63-18046

Second Printing

Printed in U.S.A. by
NOBLE OFFSET PRINTERS, INC.
NEW YORK 3, N. Y.

PREFACE

THE fortunate organization of higher studies in the University of Neuchatel [1] has, for many years, given M. Arnold Reymond the opportunity of teaching the history of science in a course followed both by the students of the *Faculté des Lettres* and those of the *Faculté des Sciences*. That portion of this course which relates to antiquity is the subject of the present publication. Its merits are so apparent and so real that it would be superfluous to insist upon them.

From the first pages of the book it can be seen with what skill M. Reymond has extricated himself from the learned controversies which the historian must have mastered in order to arrive at truths so deeply hidden to-day; with what honesty in his references, with what certainty in his choice of details, he retains, in the most simple and clear manner, whatever can effectively give the reader food for thought and help him to revive in all its depth and integrity that ancient Western civilization, the perspective of which is often spoiled and distorted by a purely literary tradition.

Many great names in the realm of science are also great names in philosophy. However, there is ground for distinction between work of a purely scientific order and speculations having a universal bearing. M. Reymond has striven to define the distinction and to keep as much as possible within the philosophic limit, so that his book, far from covering the same ground as

[1] And now at the University of Lausanne.

classical studies on ancient philosophy, particularly the excellent work of my colleague and friend, M. Léon Robin, *La Pensée Grecque*, may serve as an introduction to them. But, at the same time, being both a philosopher and a man of science, he has been able to place the technical exposition in the atmosphere best adapted to put in relief the tendency of Hellenic science, the curve of its growth, and its destiny.

The achievements of science are not to be confused with its field. The latter comprises all the questions studied by men who are called men of science, whilst the former comprises only those problems which they have succeeded in solving. The achievements of Greek science are extremely limited compared with the field which the savants of antiquity have explored. But within these limits the human mind did reach the exactitude of demonstration; it gave to truth the characteristics of certainty and security without which the appeal to truth is nothing but a mask for idleness or presumption. As M. Reymond remarks, the pretension to universal infallibility could easily find satisfaction in the primitive mentality which attributes the apparent inconsistency of natural phenomena to the fundamental caprice of invisible powers. It is quite another kind of infallibility that the Hellenic genius has apprehended, when it has established the methodology of mathematical proof.

The success of this methodology was not without its drawback. With it came an intricate connection between logic and mathematics, which was only broken by the Cartesian philosophy. This solidarity, to which we owe two masterpieces, the *Analytics* of Aristotle and the *Elements* of Euclid, made Greek science timid in face of its own conquests. As M. Reymond rightly insists, the illusory shadow of Zeno of Elea must have weighed upon the genius of Archimedes and prevented him from giving to the intellectual treatment of the infinite the positive evidence and practical fecundity which we to-day know that it implied.

On the other hand, the astronomy of position, a science which is purely mathematical, is subordinated to an astrology which seemed explanatory because it entirely filled the frames prepared by the verbalism of the Aristotelian categories.

Ancient science has in this way missed the very thing which, to us, seems the essential condition of knowledge, the connection between the mathematical and the physical, between calculation and experiment. On that depend twenty centuries of history. Rome remained totally indifferent to the purely disinterested speculative spirit which the followers of Pythagoras and Plato carried to its highest expression in mathematical research. She deliberately circumscribed the horizon of science by her anxiety for immediate utility, as is shown by an almost tragic statement by Cicero, quoted by M. Reymond. The spiritual decadence linked to the triumph of Roman imperialism, only ended with the Renaissance, when Hellenizing savants re-opened the book of exact Science at the page where the Greeks of Syracuse and Alexandria had left it unfinished.

Such considerations show clearly the utility of a work as skilfully adapted to its object as this which we have the honour of presenting to the public. Thanks to it, our men of letters will have the means of completing and rectifying their knowledge of antiquity, supporting it by an understanding of the mental substructure which will enable them at last to appreciate the order and solidity of the whole edifice. But it is addressed also to our young men of science. From lack of official institutions in harmony with a general survey of human knowledge, they are, for the most part, left in ignorance of the history of science, incapable of following the way opened up by our compatriots: Paul Tannery, Pierre Duhem, Gaston Milhaud, Pierre Boutroux, whose admirable works are so often recalled to mind by M. Reymond. The study of the Past seems to be left to lovers of phrases, to devotees of the *Verbum oratio* who can only conceive a superficial and almost grotesque representation of human nature, but whose influence, preponderant in assemblies which are governed by words, directs our education in a way contrary to the needs of our civilization and our country. The present generation suffers cruelly for not having listened to Pierre Curie beseeching *that the teaching of science should be the principal teaching in schools for boys and girls.*[1]

Better informed by their own history, the future representatives of Science will understand, and will make those around them understand, that those alone whose works witness to the sincerity of their attachment to the *Verbum ratio* are the lawful heirs of the Hellenic wisdom in its true and most truly beautiful form.

LÉON BRUNSCHVICG

[1] *Pierre Curie*, by Madame Curie. Paris, Payot, 1924, p. 98.

CONTENTS

	PAGE
PREFACE	V
INTRODUCTION	I

EGYPT AND CHALDEA

1. The Mathematical Sciences	2
2. The Astronomical Sciences	9
3. The Physical and Natural Sciences . .	15

GREEK AND ROMAN SCIENCE
PART I. HISTORICAL OUTLINE

GENERAL CHARACTERISTICS	17
CHAP. I. THE HELLENIC PERIOD, 650–300 B.C. .	21
1. Ionia and Asia Minor	22
2. Pythagoras and his School	32
3. The Eleatic School	36
4. Atomistic Tendencies	39
5. Medical Science	48
6. The Exact Sciences in the fifth and fourth centuries B.C. The Schools of Athens and Cyzicus	54
7. Aristotle. The Natural Sciences . . .	61
CHAP. II. THE ALEXANDRIAN PERIOD. From 300 B.C. to the first century of the Christian Era . .	65
1. Mathematics, Physics and Mechanics . .	66
2. Geography and Astronomy	81
3. Medicine and the Natural Sciences . .	89
CHAP. III. THE GRECO-ROMAN PERIOD. From the Christian Era to the Sixth Century A.D. . .	92
1. The Romans and Science	92
2. Greek Science in the East	95

x   SCIENCE IN GRECO-ROMAN ANTIQUITY

PART II. PRINCIPLES AND METHODS

Chap. I. The Mathematical Sciences . . . 113
  1. The Purpose and Scope of Greek Mathematics 113
  2. Arithmetic and Algebra 120
  3. The Irrational $\sqrt{2}$. The arguments of Zeno of Elea. Proportions and the Method of Exhaustion. Integral Calculus . . 127
  4. Geometrical Algebra 136
  5. The Elements of Euclid. Methods of Demonstration. Axioms and Postulates . . 150

Chap. II. Astronomy 161
  1. Meteorological Ideas 162
  2. Physical Hypotheses 164
  3. Mathematical Hypotheses 170

Chap. III. Mechanics and Physics . . . 178
  1. Technical Inventions and Physical Concepts . 179
  2. Aristotelian Dynamics 183
  3. Archimedes and Statics 192
  4. Later Developments 199

Chap. IV. The Chemical and Natural Sciences . 203
  1. Chemistry 203
  2. The Medical and Natural Sciences . . . 210

Conclusion 216

Bibliography 229

List of the Principal Works mentioned . . 237

Index 241

HISTORY OF THE SCIENCES IN GRECO-ROMAN ANTIQUITY

INTRODUCTION

EGYPT AND CHALDEA

THE information which ancient Greece has left us concerning the scientific knowledge of Oriental nations amounts to little. The traditions reported by Herodotus, Diodorus of Sicily and Strabo remain fragmentary and open to doubt.[1] The same remark applies to the explanations which geometers, such as Proclus, attempt to give in order to determine the contribution of these nations to the various branches of science. Information more direct and more reliable has been supplied in the nineteenth century by archæology and the methodical study of monuments.

The drawings and paintings which appear on the walls of temples or of tombs are valuable evidence. These drawings teach us that the Egyptians knew, for example, a practical method of drawing a hexagon, but not a pentagon. The unfinished decoration of a funeral chamber reveals an application, equally practical, of proportions and of similitude. The wall to

[1] G. Jéquier, *Histoire de la civilisation égyptienne*, 2nd edition, Payot, Paris, 1923, with a systematic bibliography of the principal works on Egyptology.

2   SCIENCE IN GRECO-ROMAN ANTIQUITY

be decorated and the image-model which is depicted on it, are in fact divided by parallel lines into the same number of squares and in each square of the wall are reproduced the forms and colours in the corresponding square of the image-model.[1] Finally the shape, aspect and construction of monuments such as the pyramids bear witness to a fairly precise practical knowledge of geometry, mechanics and astronomy. As to the information furnished by hieroglyphics and cuneiforms, it amounts to little. The only document of any importance is a manual of calculation, whose author is the scribe Ahmes, and which was probably written between the years 1700 and 1750 B.C.[2]

Thus, seeing the paucity of information available, we are reduced for the most part to conjectures concerning the scientific knowledge of the Egyptians and Chaldeans. What is certain at all events is that their knowledge was always dominated by needs of a practical or religious order.

1. THE MATHEMATICAL SCIENCES

Theoretical arithmetic was little developed amongst the Egyptians, as amongst the Chaldeans.

In practice and for reckoning they made use of abacuses the arrangement of which calls to mind the ball-frame formerly used in infants' schools.[3] As a

[1] 29 Zeuthen, *Histoire des mathématiques*, p. 5.
[2] This document (Rhind papyrus of the British Museum) has been translated into German and studied by A. Eisenlohr: *Ein mathematisches Handbuch der Alten Aegypter*, 2 vols., Leipzig, 1877. Cf. 22 Milhaud, *Nouvelles Études*, p. 58. A recent and more profound study of this document has been made by T. Eric Peet: *The Rhind Mathematical Papyrus*, The University Press of Liverpool, Hodder & Stoughton, London. (See *Isis*, VI, p. 553–7.) There exists in Moscow, if it has not been destroyed in these latter years, another important geometrical papyrus which has not yet been studied, and of which no one possesses a copy.
[3] 23 Rouse Ball, *History of Mathematics*, I, pp. 3 and 132.

INTRODUCTION

written numeration, the Egyptians used the following system. A special sign represented unity, another sign represented ten, and so on. So that, if one had to write the figure 23, it was necessary to repeat three times the sign for unity and twice that for ten.[1] This proceeding made writings singularly complicated. It was the more inconvenient because the Egyptians had not our abbreviated methods of multiplication and division. For them multiplication was reduced to a series of additions, and division to repeated subtractions. A further cause of complication in the calculations arose from the manner in which the fractions were considered. The idea of a fraction must have been evolved in the mind of man at a very early period. It was imposed upon him as soon as he knew how to measure a field, for it rarely happens that the unit chosen as a measure is contained an exact number of times in material objects. This being so, the idea of a simple fraction can be conceived in two ways. One may proceed as we do. In this case, the unit is understood in the denominator which indicates the number of subdivisions into which it is divided, while the numerator shows the sum of the parts thus obtained which one wishes to consider. To write $\frac{4}{7}$, for example, is to say that of the seven subdivisions of the unit, one only considers the sum of four of them. This being so, to add or subtract two different fractions does not present any difficulty. It is enough to reduce these fractions to the same denominator, i.e. to the same divisor of the unit, then to add or subtract the numerators and the problem is solved. But it is possible, and this is what the Egyptians did, to consider the fraction as always representing a part of the same unit. In this case the fractions will always have 1 for numerator, the denominator indicating as before the number of parts into which the unit is divided. Hence what we

[1] 22 Milhaud, *Nouvelles Etudes*, p. 51.

express as a simple fraction, $\frac{2}{29}$ for example, appeared to the Egyptians as a problem, viz., to what sum of fractions of the unit is the division of 2 by 29 equal? They showed that the sum was equal to $\frac{1}{24} + \frac{1}{58} + \frac{1}{174} + \frac{1}{232}$.

When the number to divide was greater than 2, e.g. $\frac{7}{29}$, the Egyptians resolved it in the following manner:

$$\frac{7}{29} = \frac{1}{29} + \frac{2}{29} + \frac{2}{29} + \frac{2}{29}.$$

Replacing $\frac{2}{29}$ by the value found above, one obtains finally after simplification:

$$\frac{7}{29} = \frac{1}{6} + \frac{1}{24} + \frac{1}{58} + \frac{1}{87} + \frac{1}{232}.$$

The manual of the scribe Ahmes gives a table of reduction for all fractions having 2 as numerator, and the odd numbers from 3 to 99 as denominator; i.e. $\dfrac{2}{2n+1}$ (where n may have any value from 1 to 49).[1] By what process has it been possible to compile this table? This is difficult to say, owing to want of information on this point. According to M. Zeuthen the operation was originally purely empirical, as follows:[2] Given the fraction $\frac{2}{5}$, we represent the numerator by the length $a\ b$ (Fig. 1), and the denominator by the length $a\ c$.

Fig. 1.

Now, let us take a cord, equal in length to $a\ c$, which we can fold in such a way as to get one-half, one-third, etc., of its length. If we mark off on $a\ c$ half of this cord, we reach a point beyond b, if we take one-third, we fall short of b, at the point d. There still remains a length $d\ b$, which, marked off 15 times, is equal to the whole length

[1] 9 Cantor, *Geschichte der Math.*, I, p. 25.
[2] 30 Zeuthen, *Math. Wissensch.*, p. B 19.

INTRODUCTION 5

of the cord. Then $\frac{2}{5} = \frac{1}{3} + \frac{1}{15}$. It is, however, open to question whether this process always leads to the exact results given us by the table of reduction. However that may be, the practice of expressing fractional quantities by a sum of fractions all having unity as numerator, persisted amongst the Greeks until the sixth century of our era. This practice, besides, facilitated the treatment of certain problems which, for us, lead to the solution of a numerical equation. Such is the following, propounded by Ahmes : To find a number, which, increased by its seventh, is equal to 19. The answer given : $16 + \frac{1}{2} + \frac{1}{8}$ is accurate.[1]

The tablets of Senkereh, discovered in 1854 in the library of Sardanapalus IV, give undeniable proof that the Chaldeans, besides the decimal system, used an advanced sexagesimal system based on the principle of the position value of figures.[2]

These tablets, taking sixty as unit base, give us a list of squares and cubes of which the following is an example :

1·4 (i.e. 60 + 4) is the square of 8.
1·21 (i.e. 60 + 21) is the square of 9.

More recent inscriptions even show an empty space and sometimes a special sign representing zero, when that is necessary.[3] The positional notation which characterizes our arithmetic was thus clearly known by the Chaldeans, and it is very curious, seeing its practical advantages, that it did not pass into Greco-Roman science.

How were the Chaldeans led to choose the sexagesimal division as well as the decimal system ?[4] Is it because they originally divided the year into 360 days ? Or did

[1] 29 Zeuthen, *Histoire des mathématiques*, p. 8.—6 Boyer, *Histoire des mathématiques*, p. 4.
[2] 22 Milhaud, *Nouvelles Etudes*, p. 54.
[3] 30 Zeuthen, *Math. Wissensch.*, p. B 12.
[4] *Ibid.*, p. B 13.

6 SCIENCE IN GRECO-ROMAN ANTIQUITY

they desire to have as a fundamental number, the number $2^2.3.5$ which is divisible by the majority of small numbers in constant use ? Or again, is it because the hexagon, inscribed in a circle, divides it into six equal parts ? [1]

It is very difficult to decide between these various hypotheses.

It will be seen that our knowledge of the arithmetic of the Oriental nations is very small ; the same is true concerning their geometry. According to the accepted tradition of Greek writers,[2] this science owed its birth to purely practical needs. It was the overflowing of the Nile which led the Egyptians to think of geometry, for, as soon as the inundations were over, they endeavoured to restore to each cultivator the boundaries of his fields. Hence the necessity for an exact survey. The formulæ used were, however, empirical, and were far from being always accurate. For example, to estimate the surface of a quadrilateral, the Egyptians did not attempt more than finding the product of half the sum of the opposite sides ; in order to calculate the area of a circle they used a value of π equal to 3·1604 instead of 3·1415.... They knew, however, that if the sides of a triangle are respectively 5, 4, 3, it is a right-angled triangle, and they made use of this property to erect in the field a perpendicular to a straight line. For this purpose, they used a cord divided by two knots into lengths equal to 5, 4, and 3 ; by means of pegs they made the length 4 coincide with the straight line at the extremity of which the perpendicular had to be erected (Fig. 2), then keeping taut the lengths 5 and 3, they brought them together in such a way as to join the ends.[3]

It is for this reason that the Egyptian geometricians

[1] 29 Zeuthen, *Histoire des mathématiques*, p. 7.
[2] Proclus, *Com. Euclid*, 1, p. 64, 18.
[3] 22 Milhaud, *Nouvelles Etudes*, p. 66.

INTRODUCTION

were called *harpedonaptæ*, which signifies rope-stretchers.[1] It would appear also that the Egyptians, as well as the Hindoos, had discovered, before Pythagoras, the relation between the surfaces of squares constructed on the sides of a right-angled triangle. However, the demonstration which they gave of this relation must have been purely intuitive and empirical : it probably consisted in dividing the squares so constructed into small squares, all equal, and showing the equality of the sums :

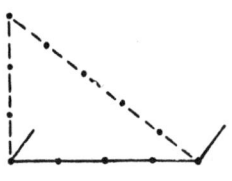

FIG. 2.

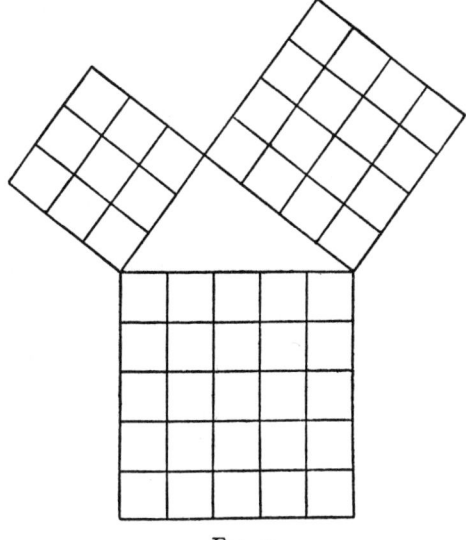

FIG. 3

$25 = 16 + 9$ (Fig. 3). This demonstration is not applicable to any right-angled triangle whatever; it neces-

[1] *Clement of Alexandria*, edit. Pottier, p. 357.

8 SCIENCE IN GRECO-ROMAN ANTIQUITY

sarily supposed the sides of the triangle to be in a certain proportion of whole numbers, 3, 4, 5, for example.[1] It may be asked whether the construction of the pyramids and temples did not require more advanced theoretical knowledge than that which we attribute to the Egyptians. M. Milhaud has clearly shown that this was not the case.[2]

Whilst the Egyptians were ignorant of the art of calculating an angle, this was the branch of mathematics which above all interested the Chaldeans. For them, indeed, the position and movements of the heavenly bodies (especially the planets) had a vital interest, since this movement influenced the destinies both of nations and individuals. So it was necessary to know how to measure exactly, at every instant, the relative positions of the planets and stars, which is impossible without the help of angles and their properties. To measure the magnitude of angles the Chaldeans, as we have seen, conceived the brilliant idea of dividing the circumference into 360 parts. Henceforward, to estimate the height of a star in the sky, it was sufficient to fix, perpendicularly to a horizontal plane, the quarter of a graduated circumference furnished with a mobile radial arm. In sighting the star by means of this radial arm, an angular displacement would be found, which corresponded to the height required. It is a curious fact, that, as we shall see, the Chaldeans had recourse to quite different methods to determine the positions of the stars. The lack of trigonometry did not impel astronomers to the direct measurement of angles.[3]

To sum up, the characteristics which distinguish Egyptian mathematics from Chaldean mathematics correspond to a difference in the practical uses to which

[1] 22 Milhaud, *Nouvelles Etudes*, p. 108.
[2] *Ibid.*, p. 75 *et seq.*
[3] 2 Bigourdan, *Astronomie*, p. 107.

this science was put. The same distinctive feature appears again in astronomy.

2. THE ASTRONOMICAL SCIENCES

In connection with the annual risings of the Nile, the Egyptians attached great importance to the exact determination of the periodic return of the seasons, and the religious festivals held in its celebration; and their observations relating to the measurement of time were far advanced.

As far back as we go in the history of Egypt, we see that the year was always divided into 12 months of 30 days each, plus 5 supplementary days; but it is probable that originally the year had only 360 days, if the following tradition, reported by Plutarch, is to be believed: [1] Saturn having secretly wedded Rhea, the Sun forbade her to give birth either during the course of a month or of a year. Hermes, the devoted servant of the goddess, played at dice with the Moon and gained from her the 72nd part of each day; thus a total of 5 supplementary days was provided during which Rhea might bring her child into the world.

The Egyptians had therefore ascertained, at a very remote epoch, that the period of 360 days for a year is too short. They recognized also that 365 days is not enough, and must be extended to $365\frac{1}{4}$ days. For if once the sun rose at the same time as Sirius on the first day of the year, the following year at the same period, Sirius would rise six hours later than the sun, and at the end of four years one day later. Thus, there would have to be $365 \times 4 = 1,460$ years before the risings of the sun and Sirius coincided again on the first day of the year. This period of 1,460 years is the celebrated Sothiac period (Sothis being the Egyptian name for Sirius) which regulated the celebration of

[1] *De Iside et Osiride*, ch. 22 (18 Maspero, *Histoire Ancienne*, p. 80).

10   SCIENCE IN GRECO-ROMAN ANTIQUITY

great religious festivals. The precision of these calculations may appear surprising at first sight, but they can be made with very simple instruments.[1]

Whilst the Egyptians were chiefly interested in the movement of the sun, the Chaldeans studied carefully the movements of the planets, believing human destinies to be bound up with these mysterious movements. Favoured by exceptional atmospheric conditions, they extended their observations very far. They quickly recognized that the planets, the sun and the moon move across practically the same region of the heavens, i.e. the zodiac or plane of the ecliptic. Therefore "as a result of their astrological ideas, the Chaldeans, instead of referring the positions of the stars or of the planets to the equator, have referred them to the mean circle of the zodiac, and this circumstance has been of great historical importance, because, when the Greeks inherited the Chaldean science, Hipparchus could thereby discover the precession of the equinoxes. It is clear that, if the system of co-ordinates by right ascensions and declinations had then prevailed, the complex law of the variations of these co-ordinates could not have been discovered." [2] It is well known that, by virtue of the precession, everything takes place as if the axis of the earth described a cone of revolution, and took a period of 26,000 years to describe it. The result of this is that the North Pole slowly changes its position in the sky, and that each year the plane of the equator cuts that of the ecliptic at a point slightly different from that at which it cut it the preceding year at the same time of the year.[3] The Greeks,

[1] 22 Milhaud, *Nouvelles Etudes*, p. 89.
[2] Paul Tannery, *La Grande Encyclopédie*, article: *Astronomie*.
[3] The following fact illustrates this progressive displacement. If the Ram occupies the foremost place in the nomenclature of our zodiacal signs, it is because, at the time when these were depicted, the sun was entering the constellation

INTRODUCTION

moreover, did not deny having borrowed from the Chaldeans the idea of the zodiac and the animal configurations which divide it into 12 regions. "They, themselves, acknowledged the fishes of the Euphrates in their sign of the Fishes. But, afterwards, they connected all these constellations with their national mythology, and thus made unrecognizable the original exotic characters which would have indicated their origin."[1] However this may be, it was by the use of the zodiacal circle that the Chaldean astronomers were able to predict with more or less exactitude the eclipses of the moon and of the sun. They noticed that the orbit described by the moon is slightly inclined to this circle, and cuts it at two points called nodes, or the head and tail of the dragon, because it is always at these two points that the eclipses of the sun or moon occur. By noting the position of these nodes with regard to the fixed stars, they were able to ascertain that these gradually moved along the zodiacal circle, and returned to their original position at the end of a certain cycle of lunations.

Having noted the succession of eclipses which were produced during the cycle, it was possible for them to predict their return. It is probable, however, that a calculation of this kind does not belong to a period earlier than the second or third century B.C., and that, before that epoch, the Babylonians were ignorant of the so-called cycle of Saros.

At first the prediction of the eclipses of the moon could be made by very simple methods, thanks to especially favourable circumstances, which return

of the Ram at the spring equinox. Now, owing to the precession of the equinoxes, it only arrives there in April. C. Flammarion, *Initiation astronomique*, Hachette, Paris, 1908, p. 147.

[1] 2 Bigourdan, *Astronomie*, p. 21.

12 SCIENCE IN GRECO-ROMAN ANTIQUITY

periodically in the course of centuries. From 755 to 432 B.C. the eclipses succeeded one another in series alternately of 5 and 6. In each series the eclipses took place every six months, and the series were separated by an interval of 17 lunations.[1] Thus it became possible to make predictions at short notice, which explains some of the inscriptions found on the cuneiform tablets.

As to the instruments of observation, we have little information. The Chaldeans certainly made use of the gnomon, which appears to be the most ancient instrument used for studying the movements of the stars, for it is everywhere mentioned before all others, whether amongst the Chinese or the Chaldeans, the Greeks or the Incas. It is, besides, a marvellously simple instrument, composed of a vertical style standing on a horizontal plane. By reproducing the movement of the sun, the extremity of the shadow projected by the style makes the division of the day possible.[2] At first sight the precision of the observations made by means of the gnomon would seem to increase with the length of the shadow, and therefore with that of the style; but in reality the shadow of the style is not very sharp because of the penumbra.[3] Further, as the length and even the direction of the shadow vary for the same hour on different days, it was necessary to have recourse to some sort of table, which gave for each month the length of the shadow at different hours.

Later on, the gnomon with the vertical style was replaced by a gnomon with the style pointing towards the pole. It was then only necessary to observe the

[1] 2 Bigourdan, *Astronomie*, p. 34.
[2] 24 Sageret, *Système*, p. 95.
[3] In order to remedy this disadvantage Facundus Novus had the idea of fixing a ball on the point of the gnomon. Pliny, xxxvi, 72.

INTRODUCTION 13

direction of the shadow in order to conveniently reckon the time.[1]

Besides the gnomon, the Chaldeans used the polos. The polos is a half sphere hollowed out in a block of stone or metal, at the bottom of which is fixed a style with its end reaching exactly to the centre of the sphere.[2] Hence the name σκάφη (boat) given by the Greeks to the polos. By this means an exact representation of the sun's movement was obtained ; " the shadow of the point of the style moves in the interior of the polos as the sun moves in the heavens, in the same direction and with the same angular velocity at each instant; it is only the sense of the motion which is reversed."[3] The hourly division of the time, represented by the meridians of the hemisphere, remains the same for all periods of the year. The shadow, in its curved path, sweeps a zone in latitude, whose breadth corresponds to the difference between the shadows projected at the summer and winter solstices.

To measure time during the night the Chaldeans at first used the clepsydra, and it is probably by means of this instrument that they divided the zodiac into 12 equal regions.[4]

Later, by combining the polos with a kind of armillary sphere, they could verify their nocturnal measurements in the following manner :[5] Imagine an open-work sphere, made of strips of metal for instance, representing the celestial sphere, and more especially the zodiacal zone with its principal constellations, and let this sphere be constructed in such a manner that it can move within the polos and be exactly adjusted to it. Suppose the zodiac to be divided into 360 degrees,

[1] 2 Bigourdan, *Astronomie*, p. 92 *et seq*.
[2] 24 Sageret, *Système*, p. 106.
[3] *Ibid*.
[4] 2 Bigourdan, *Astronomie*, p. 95.
[5] 25 Tannery, *Science hellène*, p. 84.

according to the Babylonian custom, and that, on the night of the observation, the degree occupied by the sun at the instant of its setting be known. " Then, if, at the moment of which it is desired to ascertain the time, the zodiacal stars on the eastern, western and southern horizons be observed, the stars represented on the sphere of the instrument can be brought into the same position ; the degree in which the sun is situated will then play exactly the same part as the shadow of the end of the style during the day, and its position with regard to the horary lines traced on the polos, gives the required time." The Chaldeans thus succeeded in solving, by mechanical means, problems for whose solution we have recourse to spherical trigonometry.

By dint of patient observations and in spite of the imperfection of their instruments, they succeeded in accumulating a considerable number of data, amongst others ephemerides of the sun, moon, and principal planets. The tablets of Cambyses, for example, give a list of the conjunctions of the moon with five planets, and also a list of the conjunctions of the planets with each other. The celebrated astronomer KIDINNU had calculated the synodic lunar month with astonishing accuracy, to an error of 0·4 seconds (29 days 12 hours 44 minutes 3·3 seconds instead of 2·9 seconds).[1]

However, throughout all these splendid discoveries, the distinctive features of Chaldean astronomy persist ; being calculators and traders, the Chaldeans merely sought to draw up numerical tables which would meet their astronomical needs. They did not seek, as did the Greeks, to represent geometrically the real or apparent movements which explain the variable positions of the heavenly bodies on the celestial sphere.

[1] 2 Bigourdan, *Astronomie*, p. 217.

3. THE PHYSICAL AND NATURAL SCIENCES

The technique of the manufacture of metals and even that of perfumes seems to have reached a remarkable stage of development among the Oriental peoples. The same may be said of medicine, at least, in one of its branches. For, amongst the Egyptians, the physician was required to perform two tasks of equal importance. He had, first of all, to discover the nature and if possible the name of the malevolent spirit which, by its intrusion into the body, had caused the malady; then he had to attack it, drive it away, and even destroy it. " He can only succeed in this by being a powerful magician, expert in exorcisms, skilful in manufacturing amulets. Then, with his drugs, he must fight the disorders which the presence of a strange being produces in the body; it is a question of régime and of carefully graduated remedies."[1] In the treatment of diseases, magic and incantations play therefore the principal part.

As to medicines, they were of four kinds: ointments, potions, poultices and injections. They were composed of a large number of various natural products.[2] Most of these remedies were believed to have a divine origin. The Egyptian physicians, the majority of whom belonged to the priestly caste, also used prescriptions borrowed from the Phœnicians and Syrians, or collected during their personal practice. In this manner, the experience gained was never lost and the treasure of medical science increased from generation to generation.

The whole of this medical knowledge is not to be discarded; for instance, modern science has shown that remedies composed of excrements contain ammonia, and can be advantageously used in certain diseases. Nevertheless in Egyptian or Chaldean medical science,

---
[1] 19 Maspero, *Lectures historiques*, p. 125.
[2] 18 Maspero, *Histoire Ancienne*, p. 84.

16 SCIENCE IN GRECO-ROMAN ANTIQUITY

there is nothing but prescriptions, methods, and formulas of every-day practice.[1]

In the domain of physical science, likewise, the Oriental cosmogonies reveal no sign of systematic conceptions. The Egyptian and Chaldean astronomers held that the primæval chaos became order by the effect of a divine will. The sky became a liquid mass which surrounded the earth, and rested upon the atmosphere as upon a solid foundation. The planets and all the stars floated on this celestial ocean, each sailing in its boat in the wake of Osiris. Another theory, equally widespread, represented the fixed stars as lamps suspended from the celestial vault, which a divine power lit every evening to illuminate the nights of the earth.[2]

[1] This statement perhaps needs qualification. A papyrus recently discovered by Edwin Smith and studied by J. H. Breasted (*Recueil Champollion*, 1922, p. 387-429) describes and diagnoses in order, beginning from the head, the principal diseases, indicating appropriate remedies for them, and the possible chances of recovery.

[2] 18 Maspero, *Histoire Ancienne*, p. 78.

GREEK AND ROMAN SCIENCE
PART I. HISTORICAL OUTLINE
GENERAL CHARACTERISTICS

AMONGST the problems with which Greek science confronts us, there is one which is particularly complicated, that of its birth. This has doubtless been influenced by the intimate connection which existed between the inhabitants of the countries bordering on the Ægean Sea and the East, particularly Egypt, as is shown by their many commercial transactions. The Greeks themselves are unanimous in recognizing this (legend of Cadmus, traditions reported by Herodotus, and by Proclus in his *Commentaries on Book I of Euclid*, etc.).

The question here arises in what really consists this influence of Oriental thought on Greek science? Has the latter merely received from the former a mass of empirical knowledge, or also, in some measure, the rational direction which characterizes it? The recent discoveries of Minoan civilization have further complicated this problem. The remains of this civilization seem to have survived, outside Greece and Crete, for some time after the Dorian invasions.[1] Did these remains, together with material imported from the East, form the foundation of the civilizations which

---

[1] R. von Lichtenberg, *Die aegaische Kultur*, Teubner, Leipzig, 1911. See also the complete and graphic work just published by G. Glotz: *La civilisation égéenne*, Renaissance du Livre, 1923, p. 445, *et seq.*

in the eighth century B.C. flourished in the coastal regions of Asia Minor ? It is difficult to say, for lack of historical data. But it seems probable that the characteristic rationalism of Greek science is proper to this science ;[1] in regard to the empirical and fragmentary knowledge of the East, it constitutes a veritable miracle. For the first time, the human mind conceived the possibility of establishing a limited number of principles and of deducing from them a number of truths which are their strict consequence. This achievement, without analogy in the history of humanity, is all the more astonishing because Greek science, in its first beginnings, had a precarious existence. Not having any influence upon economic life, it could only exist within the schools of philosophy, whose lot and vicissitudes it shared. It developed spasmodically in a discontinuous fashion, in different countries, according to the civilizations which sporadically arose on the borders of the Mediterranean. Its first cradle was Ionia, of necessity the intermediary between Greece and Oriental civilization, but in consequence of the political troubles which disturbed this country, Greek science was transported into Greater Greece, in the South of Italy. It was there that Pythagoras and his school established the lasting foundations of the geometrical and astronomical sciences, which the Greeks afterwards employed. We know how, even during the lifetime of Pythagoras, a revolution put an end to the school he had founded, without however compromising the existence of his doctrines. These survived partly in Greater Greece, where they inspired the subtle dialectic of Zeno of Elea ; and partly in Greece and the countries which came under Greek influence. They also helped to establish new centres of scientific life: amongst others, at Athens,

[1] 22 Milhaud, *Nouvelles Etudes*, p. 99.—25 Tannery, *Science hellène*, p. 62.—17 Loria, *Scienze esatte*, p. 5.

at Cyrene on the African coast, at Cyzicus on the borders of the Sea of Marmora.

From the fourth century B.C. onwards, Greece lost her economic and political independence. The effect of the conquests of Alexander was the transference of scientific activity to Alexandria in Egypt, and in a lesser degree, to Pergamum, in Asia Minor. When the Roman Empire was definitely established in the first century of the Christian era, Rome and Athens naturally became the fostering centres of science, on the same basis as Alexandria. Owing to the religious and political revolution achieved by Constantine, the Hellenized Orient recovered an independence and vitality, which were lacking in the Latin Occident; the sciences, nevertheless, were in peril. It was the age of decadence, or better still, as P. Tannery has put it, the age of commentators.[1]

Accordingly, the development of the Greek and Roman Sciences can be divided into three quite distinct periods:

1. A Hellenic period (from its origin to the conquests of Alexander, i.e. from 650 to 300 B.C.).

2. An Alexandrian period (from the dynasty of the Ptolemies, about 300 B.C. until the Christian era).

3. A Greco-Roman period (from the Christian era until towards the middle of the sixth century).

[1] 25 Tannery, *Science hellène*, p. 7.

CHAPTER I

THE HELLENIC PERIOD
(from 650 to 300 B.C.)

THE beginnings of this period are marked by an intimate mingling of scientific, cosmogonical and philosophical considerations. If Hegel is to be believed, these considerations would have manifested themselves in the form of a thesis, antithesis and synthesis on the problem of existence. But the historic reality does not correspond to this brilliant conception. In fact, from its first appearance, Greek philosophical thought betrayed diverse tendencies more or less opposed, which often ignored one another. It was not with one single problem that it was occupied, but rather with a number of questions more or less disconnected, concerning the origin and the purpose of the Universe. From the first there can be clearly perceived three tendencies, which persisted through the centuries unto our own times. The school called *Ionian* applied itself to external phenomena, and endeavoured to find in them the final explanation of reality. At almost the same period, the *Pythagorean* school, in the south of Italy, sought, on the contrary, this explanation in number, an abstract principle which is not directly provided by the senses. *Heraclitus*, indeed, considered that the unstable " becoming " was the very substance of reality, and that, in order to know it, recourse must be had, not to intelligence, but to intuition.

In spite of these divergences, there is, however,

amongst all these thinkers, a certain community of ideas, in that they did not clearly distinguish between spirit and matter. The natural philosophers of Ionia, like Heraclitus, attributed spiritual properties to matter, and the Pythagoreans considered numbers as having perceptible and even moral qualities. The differences of opinion, however, rapidly became more and more accentuated. The *Eleatic* school, which was inspired by the speculations of Pythagoras, tended towards idealism; whilst the new Ionian school, whose last representatives were *Leucippus* and *Democritus*, enunciated the theories of the atomic philosophy and prepared the way for materialism.

1. IONIA AND ASIA MINOR

The ancient philosophy of Ionia is often given the name of Hylozoism. Its chief characteristic is the inseparable connection between matter and life, every material element having life and reciprocally. Therefore, the discovery of the fundamental material element is sufficient to explain all reality.

Amongst the representatives of this school may be pointed out, on the one hand, Thales, Anaximander, Anaximenes, and, on the other, Heraclitus, whose ideas remain of fundamental importance to philosophy, but of little interest to the history of science.[1]

In the seventh century before our era, Miletus still enjoyed her political independence, and kept up a flourishing commercial connection with Egypt and Babylon. It was in this town that THALES lived (about 624–548 B.C.). According to tradition, he made his fortune by selling salt; but he also used other means: one year, foreseeing an abundant harvest, he

[1] 25 Tannery, *Science hellène*, pp. 52–200.—8 Burnet, *Aurore*, pp. 37–85, 145–194.—17 Loria, *Scienze esatte*, pp. 11–25.—22 a Robin, *Pensée grecque*, pp. 41–56, 86–94.

THE HELLENIC PERIOD 23

rented all the olive trees and thus made a good profit. In his capacity of merchant he seems to have travelled in Egypt and perhaps even in Chaldea. According to Herodotus (I, 75) Thales accompanied Crœsus, probably as an engineer, in his unfortunate military expedition against Pteria. Herodotus (I, 74) also attributes to him the prediction of the solar eclipse which put an end to the war between the Persians and the Lydians, and which took place in either 610, 597, or 585 B.C., this last date being the most probable. M. Bigourdan, however, believes this to be a legend,[1] as the cycle of Saros by which solar eclipses were predicted had not been established at that epoch. But in verification of this fact, ancient testimonies may be quoted, amongst them, that of Xenophanes (Diogenes Laertius, 23), and it might be explained in the following manner: as we have seen, in the seventh century B.C. the Babylonians, owing to the simpler periodicity of the eclipses of the moon at this epoch, were able to predict them without the aid of the cycle of Saros. It is quite likely that they also ventured to foretell the eclipses of the sun, and that Thales might have brought back from one of his travels their predictions, which by chance happened to be correct for the eclipse of 585 B.C.

Thales might also have brought back from his travels the Egyptian knowledge of the division of the year and of the solstices. His cosmogony likewise seems to betray an Oriental origin. The following are its outstanding features. Water is the origin of everything. Expanded by evaporation it produces air; congealed and contracted, it gives birth to the earth. The alluvial deposits at the mouth of rivers confirmed this belief in a water which could change into earth.[2] Moreover every living organism perishes when it is deprived of water.

[1] 2 Bigourdan, *Astronomie*, p. 44.
[2] 8 Burnet, *Aurore*, p. 50.

This being so, the universe is a great liquid mass, which encloses a large hemispherical bubble of air (Fig. 4). The concave surface of the bubble forms the sky, while on the plane surface, the earth, which is cylindrical, floats like a cork. The stars are boats steered by divinities; the interior of these boats is luminous, but not the exterior, so that, when the stars float on the diametral surface of the bubble, they are invisible. The eclipses are produced every time the boats of the Sun or Moon tend to overturn.

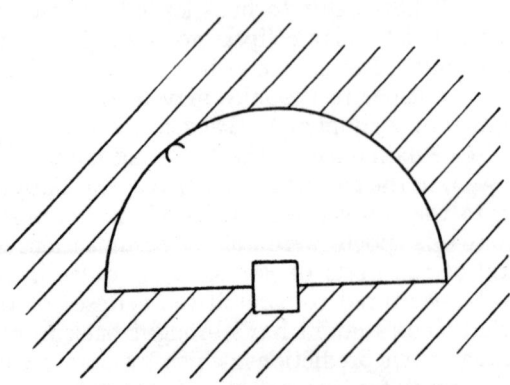

FIG. 4.

According to P. Tannery, this conception is fundamentally of Egyptian origin, but to Thales belongs the merit of having rationalized it by interpreting it according to a rudimentary natural philosophy. Thus, from the beginning, Greek thought asserted at the same time its dependence and its independence with regard to the East.

In another realm of knowledge it appears that Thales also imported into Ionia the methods of surveying in use in Egypt. Was he, however, the founder of rational geometry? It is difficult to say, although it is true

THE HELLENIC PERIOD

that the theorem of proportions by which he calculated the height of the pyramids, and the theorem of the triangle inscribed in a semi-circle are attributed to him. In arithmetic, it appears to have been Thales also who introduced into Greece the use of Egyptian fractions, the numerator of which is always equal to one.

Belonging to a younger generation, ANAXIMANDER was the disciple as well as the fellow-citizen of Thales. He was born about 610 B.C.; the date of his death is uncertain, but is generally supposed to be about 546 B.C. Anaximander wrote a treatise which contained his doctrines, and which Theophrastus certainly may have read. According to the latter, this doctrine was as follows (Diels, *Dox*, 476, 3): " Amongst those who admit one sole primary element, mobile and infinite, Anaximander of Miletus, the son of Praxiades and disciple and successor of Thales, says that the ἄπειρον is the essence and element of beings; it was, besides, he who first introduced this term of primary element, understanding by this, not the water or any other of the elements known to us, but a certain endless unlimited mass (ἄπειρον) from which were formed all the heavens and all the worlds which they have contained, etc." What is to be understood by the word ἄπειρον? Does it stand for a substance extending to infinity in space, or a substance finite in its extent, but qualitatively indeterminate? The great majority of commentators, ancient or modern, lean towards the former interpretation; at the origin of all things is a primitive matter, which extends to infinity and which we cannot perceive, since it has been transformed into derivative elements such as water, fire, etc. Teichmüller and P. Tannery consider that such a conception cannot be attributed to Anaximander, because the idea of spatial infinity only appeared later in philosophy and in science.[1] It is, moreover, in disagreement with the perception of the motion, which, according to Anaximander, brings the heavens back into the same position every 24 hours. Therefore it is in a qualitative sense that primitive matter is, not infinite, but indeterminate, that is to say, susceptible of taking manifold and varied properties. The same divergence of opinions exists as to the ideas of Anaximander on the progressive constitution of the universe.

According to J. Burnet the ἄπειρον is submitted to shocks which shake it up and down, and which, in certain regions, determine the opposition of heat and cold.[2] The heat then appears as a sphere of flames, which surrounds the cold, represented by a world whose entire surface is covered with water. Under the influence of the heat, part of the water evaporates and changes into moist air. Owing to its force of expansion, the air then penetrates the flaming sphere and divides it into rings, in which the fire is imprisoned and becomes invisible. It can, however, escape if an opening has been left by chance on the ring, when, bursting forth, it takes again its luminous consistency and forms one of the heavenly bodies we see. This being so, the eclipses of the sun or moon, and the waxing and waning of the moon are easily explained. These phenomena occur every time the openings of the solar or lunar rings become completely or partially closed up. This explanation appears, at first sight, surprisingly ingenious. Anaximander, however, may have derived the inspiration of his theory of rings from the appearance of the Milky Way, and, on the other hand, he has but extended to the heavenly bodies the explanation which he gave of lightning and thunder, namely, a fire escaping through the air contained in the clouds. "Anaximander held

[1] 25 Tannery, *Science hellène*, p. 94.
[2] 8 Burnet, *Aurore*, p. 62.

THE HELLENIC PERIOD

that thunder and lightning are caused by the wind. When it is imprisoned in a dense cloud and escapes with violence, the disruption of the cloud produces the noise, and the rent appears luminous in contrast with the darkness of the cloud " (Aetius: Diels, *Dox*, 367, 22). However this may be, Anaximander held that there were three distinct regions in which the rings were placed: the rings of the fixed stars formed the nearest region, beyond was the ring of the moon, and, further still, that of the sun.[1]

Teichmüller and Tannery admit this conception as a whole, but for them the eternal movement which animates the ἄπειρον is not an irregular disturbance, it is the movement of the diurnal rotation.[2] It is this movement, which, in the midst of primitive matter, creates the opposing forces, places in the centre of the universe the heavier elements, namely, the earth and water, then disposes around the earth the lighter elements, an envelope of air, and an envelope, lighter still, of fire. Finally, it is the centrifugal force, created by the movement of rotation, which causes the sphere of fire to burst and to divide into rings. The question of the innumerable worlds, of which Anaximander admits the existence, likewise gives rise to a divergence of interpretation which is explicable for the same reasons. The adherents to a qualitative ἄπειρον, limited in space and subject to a perpetual movement of rotation, think that by " innumerable worlds " we must understand that the actual world will be disintegrated and destroyed by the same cause (diurnal rotation) which has created it. Thus a state of chaos will be produced, from which will arise a new world, and so on. If, on the contrary, we believe ἄπειρον to

---

[1] The respective distances of the rings were fixed by sacred numbers, and not by observation.—22 *a* Robin, *Pensée grecque*, p. 49.
[2] 25 Tannery, *Science hellène*, p. 88.

be matter spatially infinite, it is more natural to admit that, in the universe, innumerable worlds can arise and develop at the same time. The word innumerable then signifies a co-existence in space, and not the simple enumeration of worlds succeeding one another in time. Thus, the cosmology of Anaximander can be consistently interpreted in two opposite ways, and, considering the texts which have been preserved, it is difficult to make a choice. The whole problem is focussed on the following question : Was it possible in the seventh century B.C. to conceive a universe, which, without being infinite in the modern sense of the term, was unlimited to such a degree that one region alone of this universe could be subject to a general movement of rotation ?

There remains to mention the views of Anaximander on the birth of living beings, for they are a very singular anticipation of evolutionary doctrines. " The first animals were produced in moisture, and were each covered with a spiny integument ; in course of time they reached dry land. When the integument burst they modified in a short time their mode of living." (Aetius : Diels, *Dox* 579, 17). " Living creatures were born from the moist element when it had been evaporated by the sun. Man in the beginning, resembled another animal, to wit, a fish." (Hippolytus : Diels, *Dox*, 560, 6).

Finally, a persistent tradition, reported by Strabo on the authority of Eratosthenes (Diels, *Vor.* 1, 12, 41) attributes to Anaximander the first geographical map. He was also supposed to have introduced into Greece the use of the gnomon and of the polos.

ANAXIMENES, the successor and associate of Anaximander, was the last representative of the School of Miletus. We do not know the exact period at which he lived, except that he was younger than Anaximander

THE HELLENIC PERIOD

and reached his " acme " [1] before 494 B.C., the year when Miletus was conquered. He produced a work, which has actually survived till the age of literary criticism.[2] His ideas are less daring but perhaps more thought out than those of his predecessor. For him, the air was the primitive boundless matter, which by condensation gave birth to earth and water, and by rarefaction to fire. It must not be forgotten that for the first cosmologists the air was always a form of vapour, darkness being another form. It was Empedocles who first discovered that the air is a distinct body, differing from vapour, and from empty space. It was he also who showed that darkness is a shadow. Anaximenes introduced several interesting theories on astronomical matters, thus justifying the esteem in which he was held by the Ancients. He considered the celestial vault, to which the stars are fixed, as solid and turning round the earth. In the interior of this vault float the sun, moon and planets, upheld by the surrounding air. In this way, the planets are distinguished from the stars for the first time in Greek astronomy. Anaximenes also supposed that dark solid bodies wander under the celestial vault. " The heavenly bodies proceed from the earth whose moisture has evaporated and, by its expansion, has formed fire; the latter rises and forms the heavenly bodies. In the region occupied by these, there are also bodies of a terrestrial nature, carried likewise by the movement of revolution " (Hippolytus : Diels, *Dox*, 561, 4). This was a fruitful conception, for it was bound to lead to the true explanation of eclipses. Indeed there was but one step needed to arrive at the supposition that the moon is one of these dark bodies, illuminated wholly or partly by the sun, according to its position, and capable of being eclipsed by the shadow of the earth.[3]

[1] Epoch of full intellectual maturity ; about the age of 40.
[2] 8 Burnet, *Aurore*, p. 77.
[3] 25 Tannery, *Science hellène*, p. 152.

The advance of the Persians in Lydia put an end to the School of Miletus, and, at the beginning of the sixth century B.C., caused an emigration of philosophical thought to Sicily and the south of Italy. The introduction of Oriental cults, amongst which the most important was that of Dionysus, caused, at the same epoch, in Greece and the Greek colonies, a religious awakening which had a profound effect on philosophical speculations.

Although belonging to a younger generation than Pythagoras and Xenophanes, HERACLITUS of Ephesus must be mentioned before them. Descended from the kings of Ephesus, Heraclitus reached maturity about the year 504 B.C. He renounced his royal rank in favour of his brother and remained all his life in Ionia, living solitary and disdainful, despising alike the men of science and the common people "who cram their bellies like cattle." This contempt was partly justifiable, for the Greeks of Ephesus lived in indolent luxury under a foreign yoke.

Primarily a theologian, Heraclitus appeals not to science but to inspiration.[1] and in his writings, he expresses himself in a sibylline manner, which caused him to be designated "obscure." His astronomical system is closely related to that of Thales. The fundamental idea of this system is the πάνταρεῖ (everything is in a perpetual state of flux). Nothing is stable, nothing is fixed. Life and death, good and evil, cold and heat, change incessantly one into the other. Nothing is either this or that, but everything is becoming. This perpetual becoming has its source in the

[1] According to the majority of commentators (amongst others, 25 Tannery, *Science hellène*, p. 186), the source of this inspiration was the divine logos. 8 Burnet (*Aurore*, p. 148 and p. 159) thinks this interpretation erroneous and based on paraphrases added by the Stoics to the original sentences of Heraclitus in handing them down to us. Logos means simply the discourse of Heraclitus in as far as it was prophetic.

THE HELLENIC PERIOD

vital fire, which is transformed into all things and which is perpetually one and many at the same time. By this, Heraclitus did not in any wise think to resolve a logical problem and to affirm the identity of contradictory propositions. This problem did not present itself to his mind, it was on the ground of experience that he based his affirmation of the union of contraries. The changes which transform fire into water, then into earth, form the up-road. The changes which inversely transform earth into water, then into fire, are called the down-road. Thus between the earth and the sky there is a perpetual exchange of effluxes following a double way, ascending and descending. From the earth and sea arise effluxes, some dry, others moist. The former are of an igneous nature, they are collected in the hollow basins which constitute the heavenly bodies, at the moment when these rise on the horizon; they then ignite to become extinguished when setting, giving a residuum of water. The damp effluxes, by their mixture with the dry ones, form our atmospheric air, which extends to the moon, whence the water falls back either as rain, or frozen in the form of snow. The various proportions of the dry and moist effluxes determine the vicissitude of days and nights, months and seasons. In winter, for example, the sun in its course is lower on the horizon, and it causes a greater evaporation of the damp layers near the earth, hence the aqueous element threatens to predominate and to completely extinguish the sun, and this is why the sun must return to the north to find there new sustenance (Cicero, *de natura deorum*, III, 14).[1]

According to Tannery, the basis of these conceptions was borrowed from Egyptian solar myths, imported into Asia Minor with the cult of Dionysus; but this is a debatable point.[2] So also is the signification to be

[1] 8 Burnet, *Aurore*, p. 177.
[2] 25 Tannery, *Science hellène*, pp. 177 and 179.

given to the great year, containing 18,000 ordinary years, at the end of which there would be a universal conflagration, and afterwards a reconstitution of the universe, periodically. This law of periodicity is quite contrary to the central idea of incessant flux affirmed by Heraclitus.[1] However this may be, Heraclitus applied to anthropology his ideas of the nature of being. To him, man was a mixture of fire, water and earth. The soul is a dry vapour which, in the waking state, is nourished by the fire spread throughout the world. In sleep, the exchange is less active, there is an encroachment of the moisture contained in the body, and this is why we lose consciousness. The same takes place in intoxication. "When a man is drunk, he is led by a young beardless boy; he stumbles, not knowing where he walks, because his soul is moist." "The dry soul is the wisest and the best" (Diels, *Vor.* I, frag. 117, 118, p. 78). Finally, when the soul changes into water or fire through the predominance of one of these elements, it leaves the body to begin once again its incessant journey above and below.[2]

2. PYTHAGORAS AND HIS SCHOOL

Amongst the thinkers who, in the sixth century B.C. left Ionia in order to escape from the Persian rule, we must first mention PYTHAGORAS, who was probably born in 572 and died in the year 500 B.C. It is not easy to reconstitute the life and doctrine of this famous man from the legends which surround them, and which for the most part were the creation of Neo-Pythagoreanism in the first centuries of the Christian era.[3] In particular the lives of Pythagoras written by

---

[1] 8 Burnet, *Aurore*, p. 180.
[2] 8 Burnet, *Aurore*, p. 175.
[3] Pythagoras, for example, kills a venomous serpent by biting it; he was seen at the same time in Crotona and Metapontum, etc.

THE HELLENIC PERIOD

Iamblichus, Porphyry, and even by Diogenes Laertius, are doubtful, but they contain much more ancient material which is worthy of belief.[1] From an authoritative source we know that Pythagoras passed the first years of his life at Samos, and that he was the son of Mnesarchus. He left Samos to escape from the tyranny of Polycrates and settled in the south of Italy at Crotona, a town already famous for its school of medicine. The travels in the East attributed to him, with the exception perhaps of the journey to Egypt, appear to have been invented later to justify his teachings.

At Crotona, Pythagoras founded a philosophical-religious school, probably after the type of the Orphic communities. Its adherents were submitted to a severe discipline; they were obliged to abstain from eating beans [2] and meat, except when they were sacrificing to the gods, for in that case, an act, which in ordinary circumstances was impiety, became an obligatory rite.[3] The Pythagoreans had, moreover, to observe not only moral rules but veritable taboos, such as " not to touch a white cock; not to sit on a quart measure; not to walk on the high roads; not to leave the mark of the pot on the ashes, when it is lifted off, but smooth the ashes."

Did the school of Pythagoras really comprise various degrees of initiation, acousmatical, mathematical and physical, with an exoteric and an esoteric teaching, jealously guarded? Or were these designations invented to explain the diversity of tendencies which

---

[1] 8 Burnet, *Aurore*, p. 94.—22 *a* Robin, *Pensée Grecque*, p. 58. On the life of Pythagoras by Iamblichus, see G. Méautis, *Recherches sur le pythagorisme*, Attinger, Neuchâtel, 1921, p. 87.
[2] For the signification of this abstinence see J. Larguier des Bancels, *Archives de psychologie*, xvii, pp. 58–68.
[3] 8 Burnet, *Aurore*, p. 106.

manifested themselves later in the Pythagorean teaching ? It is difficult to say. It appears to be incorrect also to attribute to primitive Pythagoreanism, as several historians do, a political, aristocratic and Dorian ideal, and to see in the conflict of this ideal with popular aspirations the principal cause of the fall of the school. This fall was doubtless caused by the domination which the Pythagoreans had for a time over the town, and which from its religious and moral nature must have been very tyrannical. However this may be, from the beginning of the struggle with the rich and noble Cylon, Pythagoras withdrew to Metapontum where he died soon after. His disciples remained for some time in possession of power, but overcome in the end, most of them were massacred. The survivors concentrated at Rhegium, until, with the exception of Archippus, they were forced to leave Italy. It was then that LYSIS and PHILOLAUS, whose " acme " occurred about the year 440 B.C., went to continental Greece and finally settled at Thebes. In this town they founded an important Pythagorean community to which belonged SIMMIAS and CEBES, the two Thebans introduced by Plato in the *Phædo*. Philolaus, however, appears to have returned to Italy, a little before the death of Socrates in 399 B.C. At this time the chief seat of the school was Tarentum, whence the Pythagoreans directed the opposition against Dionysius of Syracuse. To this period ARCHYTAS belongs. " He was the friend of Plato and almost realized, if he did not suggest, the ideal of a king-philosopher. He governed Tarentum for some years, and Aristoxenus tells us that he was never defeated in any battle. He was also the inventor of mathematical mechanics.[1]

Thebes and Tarentum were not the only towns in which the Pythagorean doctrine found a refuge; it flourished also in other places, amongst them Phlias in

[1] 8 Burnet, *Aurore*, p. 317.

Argolis. The doctrine of Pythagoras raises as difficult problems as does his life, for he has left no writing to enable us to distinguish his own thought from that of his disciples. We can, however, affirm that he professed belief in the transmigration of souls, for the testimony of Xenophanes is precise on this point. " One day, it is said, as he (Pythagoras) was passing by a dog which was being beaten, he cried, full of pity, ' Stop, beat no more, it is the soul of a friend; I recognized it, hearing its complaints.' " (Diels, *Vor.* I, 47, 20). On the other hand, Pythagoras was in reality a great thinker, as the testimony of Heraclitus proves. " Pythagoras, the son of Mnesarchus, extended his researches further than any other man, and choosing from certain writings, claimed as his own wisdom what was only polymathy and art of wickedness." (Diels, *Vor.* I, 80, 14).

As a thinker, Pythagoras was certainly struck by the fact that phenomena which are heterogeneous from the point of view of sensation, may nevertheless show a definite numerical relationship. Figures very different in shape may have the same surface. Musical sounds are produced according to intervals (octave, fifth, fourth) which follow a numerical law. Imbued with this idea Pythagoras extended the study of arithmetic beyond commercial needs (Stobæus, I, p. 20, 1)[1] He and his school came to the conclusion that number and its properties constitute the basis of all things. Hence, number is not a pure abstraction, it is a concrete reality, although our senses cannot directly apprehend it. Numbers have each spatial, physical and even spiritual properties, clearly defined. By their combinations they give birth to the beings and the things which we see. The contributions which the Pythagorean school made to arithmetic, geometry and astronomy were very remarkable. They definitely directed Greek

[1] Quoted by 8 Burnet, *Aurore*, p. III.

36 SCIENCE IN GRECO-ROMAN ANTIQUITY

science along rational paths. Some Pythagoreans also, for example, Philolaus and Alcmæon, carried out successful physiological and medical researches.

3. THE ELEATIC SCHOOL

XENOPHANES is generally considered the foremost representative of this school; he was born in 576 B.C. at Colophon, when this opulent city had been under the Lydians for 60 years. Driven from his native land, he travelled through Greece, criticizing the religious opinions and social customs of his time.[1] He finally settled in Sicily, but he does not seem to have stayed at Elea, although he had composed a poem in honour of this town. He died in 480 B.C. The cosmology of Xenophanes is not of great scientific interest for his aim was primarily to discredit anthropomorphic conceptions of the divinity. Being convinced that men made gods in their own image, Xenophanes affirmed the existence of a God, one, eternal, immovable who, seeing and hearing all, governs all things. This affirmation must not, however, be interpreted in the sense of a spiritualistic monotheism. The one God of Xenophanes is the heaven, the perceptible universe to which the poet attributes senses and intelligence. It is composed of two regions: the earth, flat and immovable, which extends in all directions, and the air which covers it, also illimitable. The heavenly bodies have nothing of the divine, they are incandescent clouds, similar to St. Elmo's fire; they become ignited at one end of the earth, then follow a rectilinear trajectory, and, as meteorites, bury themselves in the sands of the desert. The moist vapours of the night incessantly form new clouds, which are lit up in the morning; in this way a

---

[1] 14 Gomperz, *Penseurs* (1, p. 167), represents his life as that of a Homeric poet; but 8 Burnet, *Aurore*, p. 129, disputes this point.

new sun is born each day. Thus he explains the fact that the universe is motionless although it appears to move as a whole. By an optical illusion, easy to detect, we attribute to everything the changes which characterize particular phenomena. Thanks to the distinction which he established between the apparent and the real, Xenophanes opened the way for his disciple PARMENIDES.

According to Diogenes Laertius (IX, ch. III, 23) Parmenides reached his " acme " in the year 500 B.C.; but if we accept the somewhat doubtful indications given by Plato in his dialogue (*Parmenides*, 127 b), he was born in 516 B.C. and could not have reached his "acme" before 480 B.C. He scarcely left Elea, his native town. It was there that he received instruction from the Pythagorean Ameinias, who made a profound impression upon him. In the famous poem which he wrote, he shows us the virgins, daughters of the Sun, leading him to the dwellings guarded by avenging Justice and inhabited by the Goddess, who takes him by the hand and teaches him to distinguish between the truth which rests on the real being and the ideas suggested by appearances. The Being or Ent is what the intelligence understands and plainly identifies; the not-Being, or Nonent, is what cannot exist because of internal contradictions. The real Being is space materially extended, immovable, indivisible, uncreated and imperishable; this space is also limited and spherical, for an indefinite whole is inconceivable. The not-Being is empty space, the conception of which corresponds to nothing, since by its definition the empty space excludes all positive reality. Beside the true Being there are particular phenomena, changing and perishable. These arise from appearances and can only create ideas in our mind. In expounding these ideas, Parmenides is inspired by the cosmology of

Anaximander, complemented by that of Pythagoras. He also sets forth ideas on physiological subjects in accord with the medical science of his time. By placing in opposition the immovable and indivisible Being, and the sensible phenomena which move and divide, Parmenides raised a problem which up to modern times has been a stumbling block to philosophical reflection : what relation is there between the movement of an object and the immovable portion of space in which this movement takes place ?

This problem was clearly defined by a disciple of Parmenides named ZENO. Twenty-five years younger than his master, Zeno also dwelt at Elea where he was born in 489 B.C.; he took an important part in the direction of public affairs and meanwhile made a journey to Athens, which was recorded by Plato. According to tradition, Zeno was put to torture for having conspired against a personage who tyrannized over the town of Elea, and, rather than denounce his fellow conspirators, he cut out his tongue. Tradition also attributes to him several works : *An Interpretation of Empedocles*; *Against Philosophers*; *The Disputations*; *Treatise on Nature*. Because of the systematic manner in which he exposed and criticized the opinions of his adversaries, he was called by Aristotle the father of dialectic. (Diog. Laert., IX, 25).

As to his celebrated arguments, some have been preserved to us by Simplicius [1] and others by Aristotle (*Phys.* VI; 239 b, 9–33). The former treat of the relation of unity and plurality; the latter of the problem of motion. What was the exact meaning of these terms in the doctrine of Zeno ? Did he, by their exposition, attempt to demonstrate the impossibility of motion and of plurality ? Or did he simply desire to

[1] Ritter and Preller, *Historia philosophiae graecae*, 9th edition, Gotha, 1913, 131–4.

prove that the Pythagorean theses on discontinuity led to consequences still more absurd than the affirmations of Parmenides ? [1] This latter hypothesis would seem to be more correct, for Plato says of Zeno concerning the subject of his writing that " this is a kind of reinforcement of the argument of Parmenides against those who try to turn it into ridicule, for this reason, that, if reality be one, this argument is entangled in a mass of absurdities and contradictions. This writing argues against those who uphold plurality and gives them as much as and more than they have given ; the aim is to show that their hypothesis of multiplicity will be confused with still more absurdities than the hypothesis of unity if elaborated with sufficient care " (*Parmenides* 128 c). Whatever may have been the aim pursued by Zeno, his reasonings have an independent value, for they emphasize forcibly the difficulty of explaining logically the relations of the one and the multiple, the finite and the infinite, the mobile and the immobile.

MELISSUS (of Samos) appears to have been like Zeno the disciple of Parmenides at almost the same epoch. He affirmed in a more systematic manner than his master the unity of Being, but his views on this subject concern the history of philosophy more than that of science.

4. ATOMISTIC TENDENCIES

Both the works and the personality of EMPEDOCLES have always been a subject of discussion. The Ancients considered him either an impostor or a genius (Lucretius, I, 716). In modern times, Hegel treats him with contempt, Nietzsche admires him, and Gomperz sees in him a precursor of modern chemists. Empedocles was probably born in 490 and died in

[1] 20 Milhaud, *Phi. geom*, p. 132.—25 Tannery, *Science hellène*.—8 Burnet, *Aurore*, p. 360.

424 B.C. He scarcely left Agrigentum, his native town, except towards the end of his life, when he was forced into exile for having ardently supported democratic principles in spite of his wealth and titles of nobility. The most diverse reports of his death have been current; according to some, he threw himself voluntarily into the crater of Etna, according to others he was hanged. But it is certain that Empedocles played an active part as philosopher, physician and politician, and that he made a profound impression upon his contemporaries. He believed in his own worth. " I am for you," said he to his listeners, " as an immortal god, no longer a man ; I am honoured by all, as is just; wreathed with fillets and green coronets, I go into the neighbouring towns receiving the homage of men and women ; they follow me in thousands asking the way of deliverance. . . ." (Diels, *Vor.* I, p. 205). Despite the high opinion which Empedocles had of himself the deeds attributed to him appear to be legendary. It was not he who made healthy the marshes round Agrigentum. Still less did he protect the town against the Etesian winds, and resuscitate a woman supposed to have been dead for thirty days. These beliefs seem to have originated from certain passages in his poem which have been distorted from their original meaning.[1]

As a philosopher, Empedocles appears to have been influenced both by Pythagorism and by Parmenides. He admits with the latter that reality is a *plenum*, spherical, continuous, eternal and immobile ; but he attempts to explain the birth of motion and sensible phenomena by a method different from that of the arithmetical pluralism professed by the Pythagoreans. The universe is based on four imperishable elements, namely, earth, water, fire, and air, which Empedocles was the first to distinguish clearly from moisture and

[1] 8 Burnet, *Aurore*, p. 235.

darkness. These elements have natural attractions or repulsions for each other which cause them to combine or to separate. They float in two surrounding media, which are love and hatred. These media, although invisible to the senses, are material forces just like the ether of the physicists. They act indifferently on all bodies. Love, for example, has the effect of uniting elements whose natural affinities do not impel them to unite; hatred, on the contrary, separates the bodies which are naturally inclined to combine. The natural affinities of corporeal molecules and the combined action of hatred and love are sufficient to explain the changes and the astonishing diversity of sensible phenomena. In the beginning the four elements formed a harmonious spherical whole, entirely enveloped by love; around the universe thus constituted extended the finite medium of hatred. This latter, similar to the empty space of the Pythagoreans, at a given moment absorbed the four elements, and taking the place of love, drove the latter to the end of the world, thereby creating a veritable chaos. But this chaos did not last for ever; a movement of revolution was gradually produced in the universe, at first very slow (nine months instead of a day) then becoming more and more rapid. The central region was but little affected by this movement of universal rotation, and it was into this region where tranquillity reigns that love hastened to build up the world anew.[1] The air escapes first, but compressed by the limits of the universe, it is transformed into a hollow crystalline sphere. Fire

[1] Here we are following the current interpretation, which is also that of 25 Tannery, *Science hellène*, p. 310, but not of 8 Burnet, *Aurore*, p. 268, who thinks that according to Empedocles our actual world would be in the cycle of disorganization due to hatred, and not in the period of organization by love. This difference of opinion is of secondary importance because it does not modify the cosmological conceptions of Empedocles as a whole.

accumulates on one half of this sphere, making it luminous; the other half remains dark. This is why the earth, placed in the centre of the universe, sees the alternation of day and night. As to the sun, it is merely the image of the earth, produced by reflection. " The sun is not a fiery substance, but an image of reflected flame, similar to that which comes from water " (Diels, *Vor.* I, p. 158, 35). The light which comes from the fiery hemisphere strikes the earth, then, concentrated there, it is sent back on to this same hemisphere, where it appears to us as a luminous disk. That this really was the idea of Empedocles, Plutarch confirms by one of the characters he introduces ; " You laugh at Empedocles," said he, " because he attributes to the sun the following origin : the light of the sky after having been reflected on the surface of the earth, reflects the image of the earth again on the sky " (Diels, *Vor.* I, p. 188, 8). This conception, although at first sight curious, is very easily explained.[1] The discovery had just been made that the moon shone by reflected light, and Empedocles was naturally led to give to this theory a wider application than was legitimate.[2]

[1] 8 Burnet, *Aurore*, p. 272.
[2] It is interesting to compare the views of Empedocles with the ideas expressed by the astronomer Nordmann in his scientific romance, entitled " Einstein and the Universe." The curvature of space being constant and such that it closes upon itself like a spherical surface, one may imagine " that the rays emanating from a star, from the sun, for example, will converge at a diametrically opposite point of the Universe, after having gone round it," and that they thus form a new star. It is true, adds Nordmann, we have not yet been able to prove the existence of these phantom stars, " But what observers could not do yesterday, they will be able to do to-morrow by the help of the suggestions of the new science." One can thus foresee " the surprising and unexpected consequences of the new conceptions, which exceed in their fantastic poetry all the most romantic constructions of imaginative extrapolation. The real or at least the possible ascends

THE HELLENIC PERIOD

Beside the nature of the lunar light, Empedocles laid claim to another equally important discovery of his times, which enabled him to determine the true cause of solar eclipses. On the other hand, he professed strange opinions on the evolution of living beings. These had their birth in the following manner : the limbs, heads, arms, legs, etc., appeared separately, then they were united indiscriminately by love. Thus there came into existence cattle with human heads, and monsters with several heads. Of these strange animals only the fittest survived and perpetuated themselves by the ordinary methods of procreation. In physiology, Empedocles maintained that respiration takes place not only by the mouth, but by all the pores of the body. He also had an interesting theory of perception which has been preserved by Theophrastus. Perception is due to the contact of an element found in the organs of our senses (fire, for example, in the eye) with the same element placed outside us (Diels, *Dox*, 500, 19). By his conceptions as a whole and above all by his doctrine of the four elements, Empedocles was bound to exercise a lasting influence on medicine as well as metaphysics.

ANAXAGORAS was the contemporary of Empedocles and Leucippus. He was born in the year 500 B.C. at Clazomenae, where he possessed much property. In order to devote himself entirely to philosophy he converted his arable land into pasture for sheep, and afterwards left it entirely to his family. He then settled in Athens, where he introduced philosophical speculations,

to giddy heights, which have never been reached by the golden wings of fancy " (p. 180 *et seq.*). Empedocles would certainly have been surprised, could he have known that at the beginning of the twentieth century the theory of phantom stars, in a finite and spherical universe, would be considered as a giddy height which human imagination up to the present had never dared to attempt.

but being accused of atheism for having said that the heavenly bodies were simple material bodies, even the friendship and protection of Pericles could not save him from banishment, and he took refuge at Lampsacus, where he died in 428 B.C., honoured by all for the nobility of his character. He left a " Treatise on Nature," several fragments of which have come down to us. In this treatise he is the first to give the true explanation of the phases of the moon, likewise the first to discover the true nature of the light of the moon, and consequently he expounds the theory of eclipses. On the other hand, he considers the sun, moon, and all the stars to be burning stones, which are moved in a circle by the rotation of the ether. Unfortunately, on other points, he holds the opinions of Anaximenes, and regards the earth as a flat and concave body. As to the universe, Anaxagoras declares that it is at one and the same time infinite and animated by a movement of diurnal rotation, and in order to remove the contradiction implied by this affirmation, he admits that one part only of the universe is in motion, and, with the exception of this part, all that extends to the infinite is motionless. Motion is not inherent in matter, it is communicated to it from without by means of a subtle and intelligent fluid which is Mind or Nous (νοῦς). This is not the supreme intelligence, in the meaning which Plato and Aristotle give to this term. It is rather an organizing omniscient force, which is at the same time corporeal, personal and impersonal, and which relates more to the physical order than to the moral order (Plato: *Phaedo*, 97 c). This being so, the universe is formed as follows : The Nous puts in motion a portion of the infinite and immobile matter, then it propagates its organizing influence over a vaster and vaster region of the universe. No limit can be assigned to this influence, since on one hand the universe is indefinitely extended, and on the other hand matter is indefinitely divisible, for vacuum is incomprehensible and therefore cannot exist. "In relation to the small, there is not a least, but there is always a smaller, for it is not possible for Being to be annihilated by division. In the same way in relation to the great, there is not a greater, and it is equal to the small in plurality, and in itself each thing is at the same time great and small" (Diels, *Vor.* I, p. 314, 16). In giving these definitions, Anaxagoras was the first to bring to light one of the aspects of the mathematical infinite, which he wrongly connects with sensible phenomena. The world is a magnitude which increases beyond all limits, and matter is indefinitely divisible. Thus, according as it is indefinitely divided or indefinitely added to, the same thing may be said to be infinitely great or infinitely small. Only if matter be infinitely continuous and divisible, how can it form individual and distinct beings? Aristotle and Zeller answer this question by saying that Anaxagoras believed matter to be composed of an infinite number of elements all qualitatively different, and which the influence of the Nous had gradually grouped according to their affinities. The various groups which were formed in this manner could separate, and this explains the birth and death of phenomena. This conception was very nearly analogous to that of Democritus. Tannery judges it unacceptable as Anaxagoras expressly declared that empty space does not exist. According to Tannery the atomism taught by Anaxagoras was essentially qualitative. The infinitesimal elements of matter are of the same nature as matter taken as a whole. For example, a part of the human body, however small it may be, contains heat, cold, hairs, teeth, muscles, etc. Finite bodies do not therefore result from a mechanical mixture of atoms differing in quality, for Being, however much it is divided, remains the same qualitatively. But, if this be so, whence come the diversities which our senses reveal to us ? They result simply, answers Anaxagoras, from the fact that the Nous intensifies such or such a quality and makes it predominate over some other in the constitution of the body. This is the reason why objects which we perceive appear to us to differ from one another, although they are composed of exactly the same substance. The qualitative atomism of Anaxagoras is a remarkable effort to reconcile the unity and plurality of Being ; but it is unfortunately a hypothesis which scarcely seems susceptible of scientific verification. It had, nevertheless, a great influence on Plato and Aristotle. How, asked Anaxagoras, can qualities, which sensation shows to be irreducible (red and blue for example) mix together ? Transferring this problem to the world of ideas, Plato likewise examined in what manner ideas, which each formed an indissoluble whole, could form a group and partake of each other's properties. As to Aristotle, if he borrowed from Empedocles the theory of the four elements, under the influence of Anaxagoras, he gave them a purely qualitative signification, which persisted during the Middle Ages and which hampered the progress of physical science, as such a conception discards the use of mathematics.

About 460 B.C., LEUCIPPUS of Miletus, a disciple of Parmenides, and a contemporary of Empedocles, taught another system of atomic philosophy more scientific and more important. His ideas were taken up and developed by DEMOCRITUS of Abdera (460–370 B.C.) who, according to tradition, travelled in Egypt and as far as the Indies. Amongst the works attributed to him, several were really written either by his master or his disciples. The outstanding idea of all these works is the following : In spite of the opinion of the Eleatic school, the existence of empty space and not-Being must be admitted, for without empty space movement is inconceivable; if movement exists, empty space also exists. On the other hand, to enable particular movements to be effected everywhere, empty space must penetrate and divide Being. But this division cannot go on indefinitely or Being would be annihilated. It results therefore that all bodies must be composed of ultimate elements or atoms. From a metaphysical point of view, these atoms possess all the properties attributed by the Eleatics to Being. They exist from all eternity, they are quite complete and hence indivisible; they are absolutely simple, without any internal property which would distinguish them qualitatively from one another. However, they differ physically in form and magnitude, and this is why the natural bodies resulting from their combination appear to us so varied. The atoms have, moreover, a weight which is proportional to their magnitude. According to Burnet, this property exists in a relative sense, for it does not appear in an isolated atom. The lightness and heaviness of the atoms is only due to a whirling collective movement.[1] Under these conditions natural phenomena are easily explained. Change, birth, and death result from the combination or dissociation of atoms. Everything is done in a purely mechanical manner, and where we believe we discern a distant action, there is an intermediate medium which transmits the action. Further, to account for the perceptions of the senses, we must distinguish between the primary or objective qualities (weight, density, hardness) and the secondary qualities (colour, taste) which depend upon our manner of perception. On this atomic theory is based the explanation given by the school of Abdera of the formation and structure of the world. Unfortunately, having once postulated the whirling movement and the combination

[1] 8 Burnet, *Aurore*, p. 396.

of atoms resulting therefrom, it simply adopted the cosmological ideas of the first Ionians, without taking account of the progress made by the Pythagoreans.

The ideas of Democritus on the soul and sensation are more interesting. The soul, according to him, is composed of round, extremely tenuous atoms of an igneous nature. Because of their tenuity these atoms continually tend to escape from the body, but respiration constantly renews their number. When this weakens, there is sleep and sometimes lethargy; when it ceases altogether, death supervenes. As to sensations, these imply a direct contact with objects or emanations coming from them. For example, if we perceive bodies at great distances it is because a group of atoms keeping the shape of these bodies makes an impression on our visual organ. In a more general way, the function of thought is connected with the temperature and mobility of psychical atoms. If the soul is too hot or too cold, it makes an inaccurate representation of reality.

As a system of philosophy, atomism marks an important stage in Greek thought. By affirming the existence of empty space, and conceiving Being under the form of immutable atoms, which incessantly unite and separate, the school of Abdera reconciles the theories of Heraclitus with those of the school of Elea. Becoming is not the whole of reality but it is an important part of it. The controversy in which Greek philosophy had been involved from its beginning was thus settled and the dialectics of Plato could come to birth.

5. MEDICAL SCIENCE

Between the fifth and sixth centuries B.C. mathematics, astronomy and biology separated more and more from philosophical speculations and began to establish themselves as independent sciences. Medical science, however, had not waited until this period to

THE HELLENIC PERIOD 49

live its own life. Philosophers such as Pythagoras and Empedocles had devoted much thought to this science. Unfortunately, all the medical literature prior to the Hippocratic writings has disappeared, absorbed by these writings. We can, nevertheless, form some idea of what medical science was before Hippocrates.[1]

It had its beginnings in magic, but the priests were able to direct it into other channels and to found numerous clinics called *asclepieia* or temples of Asclepius. The one at Epidaurus, a veritable sanatorium, was celebrated for a long time. Dreams and their interpretation played a great part in the treatment given to the sick. There were also lay asclepieia equally important. The gymnasia in which a dietetic regime was imposed upon the athletes often supplanted the other establishments both religious and lay. At this time various schools arose, amongst which must be mentioned those of Cyrene, Crotona, Rhodes, and especially Cos and Cnidus, the two most celebrated. From the sixth century B.C. the Greek physicians had acquired a great reputation. Democedes (521–485 B.C.) who, after having tended Polycrates of Samos, was taken a prisoner by Darius and became his confidential councillor, bears witness to this (Herodotus, III, 125). He came from the school of Crotona, made famous by ALCMAEON, who practised the dissection of animals, and discovered the most important sensory nerves, considering them as empty canals. He explained illness as a disturbance of equilibrium between the opposing elements which constitute the body, to wit cold and heat, dryness and moisture, etc. This Pythagorean theory had consequently a great influence on pathology.[2] Nevertheless, to the schools of Cos and Cnidus belongs the honour of having established

[1] *La Grande Encyclopédie*, article *Grèce*, with bibliographical notes. See also 14 Gomperz, *Penseurs*, I, p. 291.
[2] 15 Heiberg, *Naturwiss.*, p. 11.

the Science of Medicine, thanks above all to Hippocrates, who lived at Cos in the second half of the fifth century B.C. We know little of the beginnings of these two schools and the exact causes of their celebrated rivalry. But we know that both at Cos and Cnidus the teaching comprised: (1) ordinary lessons; (2) clinical studies; (3) a practical apprenticeship. The student was initiated by a solemn oath, which was at the same time a rule of conduct for the exercise of his future vocation. He " promised to honour his master " as his parents, to aid him in all his necessities, and to instruct gratuitously his descendants if they chose the same profession as himself. Apart from these, he might only instruct in medicine his own sons and pupils bound by contract and oath. He swore to help the sick " according to his knowledge and power " and to rigidly abstain from any culpable or criminal use of therapeutic means. He must not give poison, even to those who ask it; he must not give any abortive to women, and must not practise—even where healing seems to require it—the operation of castration, which was strongly condemned by Greek sentiment. Finally he promised to abstain from all the abuses open to one in his position, especially erotic abuse towards the free or slaves of both sexes, and he pledged himself to keep inviolably all the secrets into which he might be initiated either in the exercise of his profession or outside it.[1] Other precepts still were given: the physician must observe the most scrupulous cleanliness but avoid the abuse of perfumes; he must shun all appearance of quackery, must be modest in his fees and not demand them before giving his attendance, for fear of enervating the sick and aggravating their condition, for " where is the love of humanity,

[1] Passages taken from the translation by Littré of the *Works of Hippocrates*, and quoted according to 14 Gomperz, *Penseurs*, I, p. 297.

THE HELLENIC PERIOD

is also the love of the profession." These recommendations are the more significant, because in the absence of all supervision by the State, they formed the only official rule for the practice of medicine. They are taken from a collection of writings which bear the name of Hippocrates, but were certainly not all written by him. In fact these writings form a very varied collection; they contain fragments from the school of Cnidus, which was on many points in opposition to that of Cos; in them there are also observations of the sick, noted day by day, which were never intended to be made public; and violent criticisms against supernaturalism and mystical arithmetic. For example, in a manuscript entitled: *On ancient medicine*, the author holds up to derision those who postulate arbitrarily a single primary cause and pretend to explain all maladies by heat or cold, moisture or dryness. Such a proceeding is excusable in the speculations of philosophers, but when health and life are at stake, it is inadmissible. Every substance that gives out heat possesses special properties, which act very differently on man; it is these different effects which must be known in each particular case. General theories, such as those of Empedocles, belong to philosophy, they have no value in medical science. Doubtless the physician must strive after a knowledge of nature, but in detail. This aim can never be attained by empty speculations; experience and observation of individual cases alone are fruitful. But the task is hard, most physicians are like inexperienced pilots, who know how to navigate in calm weather, but whose incapability is revealed by the tempest at the cost of a shipwreck. Fortunately, slight maladies are more common than serious ones, in which any mistake has swift and fatal consequences.[1]

The *Collection of Observations* shows us the con-

[1] 15 Heiberg, *Naturwiss.*, p. 15.

scientious physician noting every day the state of his patient, practising "his art with reflection," and hating empty hypotheses. Elsewhere, the author of the fragment entitled : *On the Sacred Disease* (epilepsy) pours scorn on those who attribute its cause to a Divinity, Hera, Poseidon or Ares. Epilepsy is not a more sacred disease than any other, for it is due to the same natural causes. All is equally human and divine in the reality which contains nothing miraculous or mythical. Mental diseases, like all others, must be treated by a suitable regimen. Together with these general considerations, the Hippocratic writings contain more definite theories, but these are ofttimes contradictory. It is difficult, in particular, to know exactly what principles of medicine were taught by HIPPOCRATES (460–350 B.C.). One thing is certain, that he, more than any other, helped to base medical science on observation and experience, and to free it from rash philosophical speculations. He was moreover a remarkable surgeon. Littré has reconstructed his doctrine as follows : Hippocrates starts from the principle that there is no other internal force in the human body but its natural heat. Hence the essential cause of diseases must be looked for in the changes of the seasons which affect the human constitution. The air also plays an important part. Diet is less important because its errors only produce individual diseases. The pathogeny of Hippocrates is purely humoral, it has its roots in the pre-Socratic philosophy and draws its inspiration from Alcmaeon. Perfect health corresponds to a perfect equilibrium in the proportion and qualities of the four radical humours : blood, phlegm or pituite, yellow bile and black bile. Illness arises from the superabundance, alteration or displacement of one of these humours. In an unhealthy state, these may collect and be expelled (there is then a crisis). If the evacuation be incomplete congestion

THE HELLENIC PERIOD 53

and gangrene result. The crisis of the disease can be foreseen, and the skill of the physician consists in giving it a favourable turn. All the followers of Hippocrates agree on the importance of prognosis and diagnosis. The urine, salts, perspiration, respiration, sleep, temperature, etc., must be examined, and also the body as a whole. " It is not difficult to recognize the state of health of a man seen naked on the palaestra." Hence there are descriptions of the progress of disease, the accuracy of which is becoming recognized more and more by modern science. For example, Littré, for a long time, was unable to identify one of the epidemics mentioned in the Hippocratic writings, which, after having affected the throat, leaves traces of paralysis. He could do so, however, when in 1860 it was recognized by English and French doctors that this results from a form of diphtheria.[1]

In therapeutics, the school of Cos seems to have recommended regimens, rather than the remedies used by the school of Cnidus, which chiefly consisted of herbal decoctions.

As to anatomy, it progressed as far as was possible at a time when only the dissection of animals was sanctioned. The Hippocratists were acquainted with the general structure of the skeleton and the heart; they distinguished between the veins (conducting channels of the blood) and the arteries which, according to them, contained air. They knew nothing of the nervous system. Hippocrates, however, places the seat of intelligence in the brain, but this knowledge, inherited from Alcmaeon, was afterwards lost and had to be re-discovered by science. The treatment of fractures and sprains was described in a manner which is remarkable, but not so surprising, when one remembers the part played by gymnastics in Greece. In surgery, the Hippocratists were not afraid to perform

[1] 15 Heiberg, *Naturwiss.*, p. 18.

trepanning, and they describe the operations with great skill. They are cautious in recommending amputations because the only means known to them for stopping the flow of blood was a red-hot iron. When surgical intervention is possible " the patient must cry out to facilitate the operation." But for the amputation of a doomed limb, it was necessary to wait until the gangrene reached a joint. From all the preceding facts we can see to what wealth and precision of knowledge the Hippocratic writings bear witness. From ancient times they have been the subject of many commentaries the most important, the greater part of which is unfortunately lost, being that of Galen (second century A.D.).

Amongst the immediate successors of Hippocrates must be mentioned PRAXAGORAS of Cos and DIOCLES of Carystus. The latter has left precise and detailed prescriptions of hygiene to be followed from morning to evening, according to the seasons. However, the methods recommended by Hippocrates and his disciples were far from gaining universal adherence. The votive tablets found at Epidaurus betray a totally different mentality by the accounts of cures which they give. A woman, for example, remained pregnant for five years, then after a sojourn in the temple, gave birth to a boy, who by himself bathed in the stream and then began to frolic round his mother.

6. THE EXACT SCIENCES IN THE FIFTH AND FOURTH CENTURIES B.C. THE SCHOOLS OF ATHENS AND CYZICUS

The mathematical and astronomical writings of this period have not been preserved, but we can reconstruct them in some measure from the testimony of subsequent writers. Arithmetical researches were carried on along the mystical path opened up by the Pythagoreans, but did not attain to any remarkable results.

THE HELLENIC PERIOD

Geometry, on the contrary, made rapid progress. THEODORUS of Cyrene enunciated the problem of the incommensurables $\sqrt{3}$, $\sqrt{5}$, etc., up to $\sqrt{17}$ (Plato, *Theaetetus*, 147, D). Three problems especially attracted attention, for although they present themselves as the natural generalization from simple geometrical constructions, yet they cannot be directly solved by the means of the rule and compass. These three problems, famous in the history of mathematics, are: the trisection of the angle, the quadrature of the circle, the duplication of the cube.[1] They gave rise to numerous and fruitful investigations, and gradually led to the theory of conic sections. The primary impulse was given by the sophists. HIPPIAS of Elis first discovered the curve called the quadratrix. This curve (Fig. 5) is obtained by the intersection of the moving radius of a circle and a straight line which

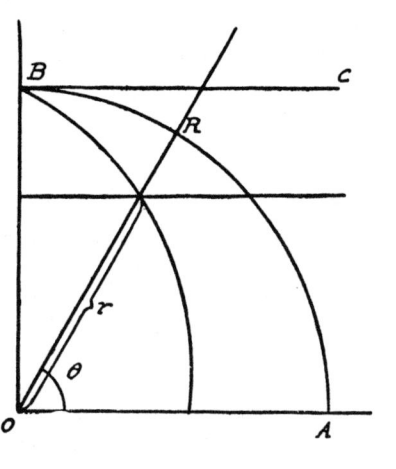

FIG. 5.

---

[1] The duplication of the cube is also called the Deliac problem. Apollo, having been consulted about the plague which ravaged Athens in 430 B.C., directed that, in order to end it, the volume of the altar of Delos, which was cubical, should be doubled. The Athenians thought to do this by simply doubling the sides of the altar; but, the scourge having redoubled, they recognized their error and applied to Plato. *Aristotelis opera*, IV, p. 209, *scholies de Philipon aux Analytiques postérieures*.

moves parallel to itself from BC to OA in the same time as the radius moves from OB to OA. The curve can be constructed by successive divisions of the arc BA and the straight line BO. This being done, it is enough to divide BO into three parts, to obtain the trisection sought. (Proclus, *Comm. Eucl.* I, p. 356, II and p. 272, 7; Pappus, I, p. 253). From an analytical point of view the equation of the quadratrix is the natural result of the following equation in which r is the radius vector of the quadratrix, a the radius of the circle, and θ the angle AOR.

We have $\dfrac{\theta}{\frac{\pi}{2}} = \dfrac{r \sin \theta}{a}$

hence $\pi r = 2 a \theta \operatorname{cosec} \theta$

The authenticity of the discovery of Hippias has often been disputed; P. Tannery, however, after detailed discussion, upholds it.[1]

Another sophist, ANTIPHON, likens the ultimate elements of the curved line to those of the straight line, and he attempts to solve problems by regarding the circle as the limit of a polygon with an infinite number of sides. BRYSON of Heraclea takes this conception and completes it by considering at the same time inscribed and circumscribed polygons. But these two sophists appear to have postulated that there is no real difference between the straight line and the curve (Simplicius: Diels, *Vor.* II, 594) and for this reason their solutions, which might have been a guiding light, remain doubtful, the more so because they bring in the notion of infinity. The disputations aroused by this subject became so popular that Aristophanes directly alludes to them (*Birds*, act II, scene vi). "These, said the astronomer Meton, are instruments

---

[1] 28 Tannery, *Mem. scientifiques*, II, p. 1.

for measuring the air. For you must know that the air is formed like an oven. This is why applying the top of this curved rule, then placing the compass, I shall use a straight rule and I shall take my dimensions so well that I shall make a squared circle."
This METON, whom Aristophanes introduces, seems to have been a good astronomer. He rediscovered the so-called cycle of Saros, which henceforward bore his name, and which helped to reform the calendar and fix religious rites. A short time after the sophists, there appeared the works of the schools of Athens and Cnidus, which were so closely united that it is difficult to separate them. According to tradition Hippocrates, Plato, and Theaetetus belong to the school of Athens, whilst Eudoxus, Menaechmus and Aristo represent that of Cnidus.

HIPPOCRATES of Chios was born in 470 B.C. Despoiled of his wealth by the Athenian customs, according to Eudemus, by pirates, according to Philoponus (Diels, *Vor.* I, p. 231, 27, 30) he came to Athens to beg for justice and the recovery of his property. Having been unable to gain his cause, he devoted himself to philosophy and opened a school of geometry. He was the first to compile a treatise of geometry, thus breaking away from the Pythagorean tradition, which kept secret all mathematical knowledge ; hereby he provided a solid basis for instruction and foreshadowed the *Elements* of Euclid. He also introduced the use of letters to indicate lines and figures, and it was really he who created the geometry of the circle by means of the two following propositions : Circles are to one another in the ratio of the squares of their diameters. Similar segments are to one another in the ratio of the squares of their chords.

Hippocrates also recognized that the duplication of the cube leads to the investigation of mean proportionals:

$$\frac{a}{x} = \frac{x}{y} = \frac{y}{b}$$

Then we have $x^2 = ay$; $y^2 = xb$, hence $x^4 = a^2xb$ and $x^3 = a^2b$. Now if we put $b = 2a$ we obtain $x^3 = 2a^3$, which is the solution required.

The quadrature of the circle is, as we know, an insoluble geometrical problem. In attempting to solve it, Hippocrates was led to several interesting discoveries on lunes. He found, for example, that the lune AECD (Fig. 6) is equal to half the right-

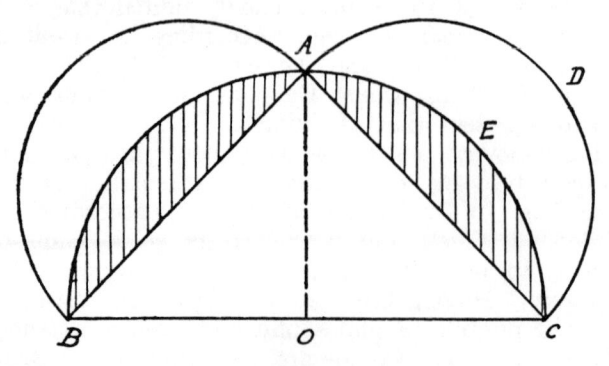

FIG. 6.

angled triangle ACB. In order to prove it, it is sufficient to notice that the semi-circle constructed on the hypotenuse BC is equal in area to the two semi-circles constructed on the sides BA and AC which, by hypothesis, are equal. If we take away the common parts of the semi-circles (small and large) we obtain the required equality.[1] Having thus demonstrated that a surface bounded by curvi-linear elements is equal to a surface limited by straight lines, Hippocrates thought it was possible to find a square equal to a

[1] 23 Rouse Ball, *History of Mathematics*, I, p. 42.

THE HELLENIC PERIOD

circle. Without labouring the point, we see how fruitful the work of this geometrician was.

ARCHYTAS of Tarentum followed it up in the duplication of the cube; he pointed out a very elegant method of discovering the mean proportionals, a method which implies a very clear understanding of "geometrical loci." According to Archytas the two mean proportionals sought are obtained by the intersection of the three following surfaces:

the cylinder $x^2 + y^2 = ax$

the cone $x^2 + y^2 + z^2 = \dfrac{a^2}{b^2} x^2$

the tore or anchor-ring $(x^2 + y^2 + z^2)^2 = a^2(x^2 + y^2)$ this latter being produced by the revolution of a circle around one of its tangents.[1]

As for PLATO (427–347 B.C.) we know the value he attached to mathematics. He borrowed from it the basis of his idealism, since mathematical demonstration cannot be based upon the observations of sensible phenomena, for Nature displays only imperfect figures. On the other hand, this demonstration could not be arbitrarily created by the mind. There exists therefore beyond the realm of sensible perception a realm of ideas of which our minds gradually become aware. Thus scepticism and sensualism are checked. Without making any real discoveries, Plato has defined the conditions of mathematical research. He insists on the necessity of reducing axioms and definitions to the smallest number possible. He distinguishes between the analytical method by which one can ascertain if the problem be solvable or not, and the synthetic method by which the solutions are worked out. In this way Plato rendered invaluable service as much in the research of primary propositions as in the construction of geometrical figures. His advice led to a

[1] 28 Tannery, *Mém. sci.*, II, p. 19.

revision of the treatise of geometry, written by Hippocrates. This revision was made first by LEON, and then by THEUDIUS of Magnesia, both pupils of the Academy (Proclus, *Comm. Eucl.* I, pp. 66, 20; 67, 12). The trend given by Plato to astronomy was no less important. His harmonious vision of the world impelled him to the opinion that the irregular movements of the planets were unreal; preserving the Pythagorean axiom of circular movement, he assigned to astronomy the task of finding a combination of circular movements which would account for the apparent irregularity of the motion of the planets (σώζ ειν τά φαιν μενα).

EUDOXUS of Cnidus, a contemporary of Plato, was a great geometrician as well as an astronomer. Born in 408 B.C., he studied under Archytas at Tarentum, then he settled with his disciples at Cyzicus, which he left for a time to live in Athens. He discovered almost the whole of the contents of Book V of Euclid, on proportions, and obtained these results by extending the notion of proportionality so as to include all rational and irrational magnitudes. He postulates that $\frac{a}{b} = \frac{c}{d}$ if $ma \gtreqless nb$ at the same time as $mc \gtreqless nd$ (m and n being numbers chosen arbitrarily and a, b, c, d, any magnitudes). On these foundations he established the basis of the method of exhaustion, so brilliantly developed by Archimedes, and which has for its complement the reduction to absurdity. To conform to the outline of astronomy sketched by Plato, he conceived a system of homocentric spheres, the essential features of which were conserved by Aristotle. It was Eudoxus also who compiled the catalogue of stars, used in the third century B.C. by Aratus in his poetic description of the starry sky; and it was he who estimated the circumference of the earth to be 400,000 stadia, a value which was accepted by Aristotle. His

THE HELLENIC PERIOD 61

disciple Menaechmus was equally remarkable. The tutor of Alexander the Great, he replied to a question of his royal pupil by saying that there are no royal roads in geometry.[1] Following the suggestions of Archytas, he resolved the problem of the duplication of the cube by finding the point of intersection of either the two parabolas $x^2 = ay$, $y^2 = 2ax$, or the parabola $x^2 = ay$ and the hyperbola $xy = 2a^2$. These equations result directly from the mean proportionals enunciated by Archytas and Hippocrates.

$$\frac{a}{x} = \frac{x}{y} = \frac{y}{2a}$$

Menaechmus may have shown besides that these curves can be obtained by the intersection of a plane and a cone of revolution, and thus opened up the way for the theory of conic sections.

7. ARISTOTLE AND THE PERIPATETIC SCHOOL. THE NATURAL SCIENCES

ARISTOTLE (384–322 B.C.) directed the study of science into new paths. The son of a physician, he was as much interested in natural science and inductive methods as in metaphysics and exact science. He was at first a disciple of Plato, but he left the Academy after the death of his master. The writings he has left are valuable and varied. The greater part have come down to us in the form of notes written for an oral exposition, and they constitute a veritable encyclopædia of the knowledge of the period. But Aristotle not only collected, systematized, and discussed the opinions of his predecessors and contemporaries, he created entirely new systems such as logic, morphology, and biological classifications. It must be noted, however, that although he had sufficient mastery of elementary mathematics to use them as illustrations of his logic,

[1] This saying is also attributed to Euclid,

he does not appear to have understood the interest of higher mathematics. The ideas of function, and of geometrical loci were unknown to him; on this point he was inferior to Plato.[1] With the help of the astronomer Calippus, Aristotle attempted to perfect the system of Eudoxus by introducing compensating spheres so as to give solidarity to the movements of the planets and of the celestial vault. He was also interested in meteorological phenomena. In his eyes it is heat which plays the most important part; it contributes to the formation of comets, the Milky Way, clouds, winds, etc. The rainbow is only a phenomena of reflection, the droplets of the cloud acting as mirrors to the sunlight. (*Meteor.*, Bk. III, ch. iv.; 373 to 32.) Aristotle approaches physics as a theorist and a metaphysician; he discusses carefully ideas of place, motion, etc., but very often interprets phenomena erroneously, although he was on the point of discovering specific gravity. With Plato, he adds to the four known elements a fifth, the quintessence. By his ideas, he has, up to a certain point, impeded the progress of physics; on the other hand he exercised a happy influence on the evolution of alchemy and consequently of chemistry. The collection of writings entitled *Problems* shows us the extent and variety of the instruction which was given in the Peripatetic School, for it deals with medicine, physiology, mathematics, optics, music, philology, etc. In this collection the *mechanical problems* are particularly remarkable, because side by side with palpable errors there are glimpses of the most important laws of mechanics (the principle of virtual velocities, parallelogram of forces, law of inertia, use of tackle). The influence of the investigations of Archytas can be seen here.[2] But, as we have already remarked, Aristotle was, before all,

[1] 21 Milhaud, *Études*, p. 101 et seq.
[2] 15 Heiberg, *Naturwiss.*, p. 35.

THE HELLENIC PERIOD 63

a biologist. It may even be maintained that his system of logic, in so far as it deals with the classification of the real, is fundamentally biological.[1] It is especially in the natural sciences that Aristotle displays his predominant qualities, creative genius, power of observation, faculty of discovering and comprehending analogies, finalistic interpretation of phenomena. Not content with co-ordinating and explaining the work of his predecessors, Aristotle was the creator of scientific zoology and comparative anatomy. He classified animals with remarkable accuracy, placing, for example, the whale amongst mammals, contrary to the current opinion of his time. His two works entitled *De partibus animalium* and *De generatione animalium* abound in observations and analogical reasonings of great exactitude. This is all the more surprising in view of the fact that Aristotle had none of the modern scientific apparatus, the microscope in particular. Such results are only obtained by dint of patience and ingenuity. Aristotle drew his information from fishermen, hunters, shepherds, etc., but he checked it carefully. He observed, analysed, and verified. By a method fundamentally inductive and empirical, he purposely, in this branch of science, puts aside philosophical speculation. Doubtless, he sometimes drew too hasty conclusions, and misunderstood the discoveries of his predecessors, especially in medical science; but in general he has the great merit of taking into consideration the opinions of all those who preceded him, and thus became the creator of the historical method.

His work was carried on by his disciples. THEOPHRASTUS, whose characters were imitated by La Bruyère, has left us a very valuable book containing the opinions of the ancient natural philosophers. MENON wrote the history of medicine; EUDEMUS, that of astronomy and mathematics; ARISTOXENUS,

[1] 7 L. Brunschvicg, *Les Etapes*, p. 72.

that of music. The work of Theophrastus is of special importance, not only for the information it contains, but also for its criticisms. Besides natural philosophy, it comprises a treatise on the sensations, and another on botany, both full of accurate and extensive observations. According to Heiberg, the most praiseworthy result of the knowledge and methods of the Aristotelian school in zoology and botany, was the description and classification of the hitherto unknown specimens of fauna and flora brought back from the expedition of Alexander the Great to India.[1]

In another realm of science, the ethnographical descriptions of ARISTOBULUS, the geographical description of the southern coast by NEARCHUS, the systematic treatise on geography by DICEARCHUS, a disciple of Theophrastus and an author much esteemed by Cicero, are all worthy of mention. The two short writings of AUTOLYCUS (spherical geometry applied to astronomy) are noteworthy as being the most ancient works on exact sciences which have come down to us. HERACLIDES of Pontus, the friend and contemporary of Aristotle, also studied Astronomy; he invented an ingenious heliocentric system, and contrary to the opinion of Aristotle, maintained the infinity of the universe.[2] STRATO of Lampsacus is renowned for his works on physics; he opposed Democritus' theory of empty continuous space, although he admits, on the ground of experience, the existence of small empty spaces distributed discontinuously in the interior of bodies.[3]

[1] 15 Heiberg, *Naturwiss.*, p. 38.
[2] The ideas of Heraclides of Pontus have been preserved by the Jew Chalcidius, who in the fourth century of our era wrote a commentary on the " Timaeus " of Plato. Doublet, *Histoire de l'astronomie*, p. 126.
[3] G. Rodier, *La Physique de Straton de Lampsaque*, Alcan, 1890.

CHAPTER II

THE ALEXANDRIAN PERIOD

(from 300 B.C. to the first century of the Christian Era)

IF the conquests of Alexander the Great caused Greek language and science to penetrate into the East, they also brought about an upheaval of existing conditions. Greece lost her creative originality at the same time as her political autonomy. Athens certainly remained the seat of the philosophical schools, but in reality other towns, foremost amongst them Alexandria, became the centres of intellectual life. This now changed its character; instead of, as in the past, spreading through small democratic states, it concentrated in the capitals of the kingdoms which arose on the ruins of Alexander's empire, and hence was confined to smaller and smaller circles, for in spite of its diffusion, the Greek language, with its characteristic syntax and vocabulary, remained an unknown tongue to the masses of Asia Minor and Egypt. The classical works of Greece could only be appreciated by the chosen few. This state of affairs was unfavourable for literary and philosophical production. The latter, when it is intended for only a small circle of readers, is no longer animated by popular inspiration, and loses itself in subtlety, affectation and erudition.[1] But for the sciences properly so called, these conditions were very advantageous. Owing to the diffusion of Greek culture throughout the eastern littoral of the Mediterranean, specialists were sure to meet with savants

[1] 15 Heiberg, *Naturwiss.*, p. 42.

capable of understanding them; thanks to the munificence of princes, they had at their disposal the necessary resources for their work, and the wise administration of the kingdom secured to them the peace of mind needful for their meditations. Such peace and material independence could not be offered by the little democratic states of Greece, always a prey to revolutions.

The Ptolemaic dynasty is especially noteworthy for its intelligent initiative in establishing Alexandria as the new and indisputable centre of Hellenic culture. The founder of this dynasty summoned to him Demetrius of Phalerus and Strato of Lampsacus, both representatives of science and the Aristotelian tradition; but it was his son Ptolemy II (Philadelphus), who, like the American millionaires of to-day, founded a museum where savants were generously supported on the sole condition of furthering science. He also established two great libraries of which Aristotle's works formed the nucleus, and which 50 years after their foundation, contained more than 600,000 manuscripts. In addition to this there was an active trade in manuscripts, favoured by the fact that Egypt possessed the monopoly of papyrus. Thanks to these exceptional conditions, Alexandria quickly became the refuge of students and professors, and even kept in touch with foreign savants. Thus the sciences in all departments made rapid progress, and reached their zenith in the third century B.C.

1. MATHEMATICS, PHYSICS, AND MECHANICS

The mathematics of this period are represented by three great names, which dominate antiquity: Euclid, Archimedes, and Apollonius. Of EUCLID (330–270 B.C.) we know little except that he was called by Ptolemy Soter to teach mathematics in Alexandria. It was there that he wrote the *Elements* which made him famous, and which, translated almost literally,

THE ALEXANDRIAN PERIOD 67

have been used in English schools until these latter years. The fame of Euclid was so great that already in the Middle Ages his existence was doubted. According to the commentators of this period, the name of Euclid does not belong to a real person but to the book itself of the *Elements*, and signifies the key of geometry ($ὔκλι$ = key, $δις$ = geometry). This hypothesis, it is unnecessary to state, is more ingenious than well-founded.[1] Doubtless the *Elements* were not entirely the work of Euclid. He borrowed largely from his predecessors, but to him belongs indisputably the merit of having developed and co-ordinated into a faultless logic all the geometrical work accomplished before him. He has brought into relief the essentially rational character of geometry, and has shown that, certain principles being postulated, the sequence of mathematical propositions unfolds itself in an irresistible manner. His method is synthetic, proceeding from the simple to the complex, i.e. starting from the most elementary figures to reach the most complicated.[2] Modern analysis proceeds in a different manner. For example, to study the curves of the second degree, it begins by assuming the general equation of conics, then by successive limitations determines the circle, ellipse, parabola, etc.

The *Elements* comprise thirteen books, each of which is prefaced by definitions of the meaning, use and limits of the concepts employed. The first book also contains five postulates and five axioms which, added to the definitions, are intended to secure the logical construction of the whole edifice. In this anxiety to distinguish rigorously the nature of the fundamental propositions, we see the effect of the Platonic investigations on the foundations of mathematics. This order, adopted by Euclid, has been often criticized even by the

[1] 23 Rouse Ball, *History of Mathematics*, I, p. 55.
[2] 29 Zeuthen, *Histoire des mathématiques*, p. 93.

Ancients, but modern researches have justified it. Even the famous postulate concerning parallels has been recognized for what it was in Euclid's conception, i.e. a proposition which establishes the existence of a point of intersection between two straight lines, if the sum of the interior angles formed by these lines and a line which cuts them be less than π. The four other postulates are for the purpose of establishing the existence and unity of the elements needed for geometrical constructions since these cannot be rigorously demonstrated. The only purpose of the axioms is to set forth as briefly and completely as possible the conditions of equality and inequality of geometrical magnitudes. These foundations once established, the geometrical edifice can be constructed theorem by theorem without any appeal to intuition.

The books which form the *Elements* are divided according to their contents as follows : I, straight lines, triangles, parallelograms, the theorem of Pythagoras ; II, geometrical algebra ; III the circle, angles ; IV, inscribed and circumscribed polygons. These four books were certainly borrowed from the Pythagorean teaching, for they avoid the use of proportions even when it would be most natural.[1] Book V, which treats of proportions, is entirely inspired by the works of Eudoxus. Book VI treats of the similitude of figures. Books VII–IX make use of the works of Theaetetus and treat of rational numbers, progressions, and continuous proportions. As to Book X (incommensurable quantities) it appears to be entirely the work of Euclid. In dealing with these questions, he uses the graphical method, which consists in representing numbers by lines, and has the advantage of providing demonstrations applicable to all numbers, rational or irrational. Books XI–XIII treat of geometry in space and are inspired by Pythagoras and Plato ; they are less finished

[1] 26 Tannery, *Géom. grecque*, p. 98.

THE ALEXANDRIAN PERIOD 69

than the others, having been left in the experimental stage. For example, congruency and symmetry are not clearly distinguished and in the chain of proofs there is sometimes a break. As a whole the *Elements* display faults of method and detail which we shall have to examine later, but they remain nevertheless an admirable work, whose solidity and success have been proved by the succession of editions through the centuries from antiquity to the Middle Ages and from the Renaissance to our own times.[1]

Besides the *Elements*, Euclid has left a collection of *Data*, the aim of which was to facilitate the analytical study of theorems. The contents of this work are the same as that of the first six books of the *Elements*, but the enunciation of the propositions is stated in the form of conditions according to which a geometrical figure is given or rather determined. For example, " if two lines enclose a given space and form with each other a given angle, and if their sum be given, then each of these lines will be given " (prop. 85). A her collection, the *Porisms*, had a similar purpose; it showed what figures could be constructed, certain conditions being given. This work is unfortunately lost; several savants have attempted to reconstruct it from some imperfect texts of Pappus, but all these attempts (including that of Chasles) have been unsuccessful.[2] Two other works also have been lost. The first treats of *Surfaces as Geometrical Loci*; the second, inspired by the works of Menaechmus and Aristo, gave the *Elements of Conic Sections*. The latter was soon supplanted by the works of Apollonius, but it has been possible to partially reconstruct it. No vestige remains of the work entitled *Fallacies*. We

[1] For the history of these editions, see 17 Loria, *Scienze esatte*, p. 190 *et seq.*, and 6 Boyer, *Histoire des Mathématiques*, p. 29.
[2] 17 Loria, *Scienze esatte*, p. 259 *et seq.*

can only suppose it to have been modelled after the type of the "Sophistical arguments" of Aristotle, and to have contained historical comments of great interest. Another dissertation, of which only the Arabic version has come down to us, entitled the *Division of Figures*, shows how triangles, quadrilaterals, and circles can be divided into equal parts, or according to a certain ratio.[1]

Finally Euclid composed books on optics (or perspective), astronomy and mathematical acoustics, all with a view to teaching. By his didactic methods, Euclid differs essentially from Archimedes, whose creative genius ranks him amongst the greatest mathematicians of all times.

ARCHIMEDES (257–212 B.C.) was born at Syracuse,[2] and was on intimate terms with, if not related to, King Hiero. It was to Gelo, the son of Hiero, that he addressed the curious problem of the Arenarius, relating to the number of grains of sand which could be contained in the universe. In spite of the advantages offered by Alexandria, he preferred to live in his own country, to which he was much attached. In his writings, for instance, he uses the local dialect rather than the common speech, thus showing his patriotism and independence of character. It was especially during the siege of Syracuse that he applied his talents to the service of his country. By his wonderful inventions, he held in check the Roman armies and fleet, commanded by Marcellus. Polybius (bk. VIII, fgmt. iv), Livy (bk. XXIV, ch. 34), and Plutarch have left us an account of these inventions, but they pass over in silence the burning of the ships by means of

[1] 15 Heiberg, *Naturwiss.*, p. 50.
[2] For a critical study of the life and works of Archimedes, consult P. ver Eecke, *Les Œuvres complètes d'Archimède*, Paris, 1921; T. L. Heath, *The Works of Archimedes*, Cambridge, 1897.

a circular arrangement of mirrors. This feat was related for the first time by Lucius of Samosatus in the second century, so it is open to question, although Buffon has demonstrated its physical possibility. It is well known how at the fall of Syracuse, Archimedes was brutally slain by a soldier, contrary to the express desire of Marcellus, and how his tomb was discovered by Cicero many years after (*Tusculanes*, Bk. V, ch. 23). According to his own testimony (Heiberg edition, II, p. 248, 8) Archimedes was initiated to astronomy by his father Pheidias; he afterwards had Conon as his friend and fellow-student, and showed himself unrivalled in the construction of astronomical instruments. He constructed two planetaria, which were taken to Rome after the fall of Syracuse. One was placed in the temple of Victory, the other was preserved by the family of Marcellus, and was admired by Cicero, who speaks of it in the following terms: "What is most to be admired in the invention of Archimedes is that he was able with a single motor to reproduce all the unequal and different movements of the heavenly bodies" (*Repub.*, I, ch. 14). In another field, the pursuit of astronomy certainly led Archimedes to the study of catoptrics (laws of reflection), and to the creation of an ingenious system of numeration by which numbers of any desired magnitude can be expressed. After having benefited by his father's teaching, Archimedes, as Diodorus of Sicily relates, must have sojourned for some time in Egypt, or he would not have brought out his works in Alexandria, dedicating them to Eratosthenes, Conon and Dositheus, who lived in that city. During that sojourn he must have had some painful experiences with certain pedantic professors, for speaking of some problems propounded by Conon, the solution of which was impossible, he says this: " Those who pretend to have discovered them all, without producing any proof, are convicted of imposture since they boast of having found a demonstration which is in fact impossible" (Heiberg edition, II, p. 5). It was likewise in Egypt, if Diodorus of Sicily is to be believed, that Archimedes discovered the screw which bears his name, called also a snail or spiral pump. This pump consists of a tube open at both ends and twisted like a corkscrew. When inclined to the vertical and rotated on its axis, it raises the water in which its lower extremity is immersed. It is doubtful, however, whether such an apparatus had not been used in Egypt before the time of Archimedes. Similarly it is not known exactly by what means Archimedes launched the huge ship which Hiero had had built, and which the Syracusans could not move from the slipway (Proclus, *Comm., Eucl.*, I, p. 63, 19). According to Plutarch the machinery used was composed of cords and pulleys, but the use of tackle had been known from the time of Archytas. It is more probable that it was an endless screw, working a system of toothed wheels.[1] However this may be, it was through meditating on the construction of these engines that Archimedes was led to formulate the exact laws of mechanics. The task which he assigns to this science, namely, "to move a given weight by a given force," is only the theoretical translation of the famous saying, " Give me but a place to stand on and I will move the earth," which he uttered at the time of the launching of the vessel, the difficulties of which have been referred to. For this reason it is very likely that the writings by which Archimedes established the basis of rational mechanics (at least as far as statics is concerned) belong to the first years of his scientific activity. Perhaps it was also at this time that he discovered the infinitesimal method of integration, based on mechanics, which he used together

[1] Ver Eecke, work quoted, p. xiii.

THE ALEXANDRIAN PERIOD

with the method of exhaustion to determine surfaces and volumes. Of his works, we only possess the following : *On the Sphere and Cylinder*, an enunciation of five postulates, which, in the absence of any consideration of mathematical infinity, allow of the demonstration of the problems proposed : area of the sphere equal to that of four great circles ; ratio of the surface and volume of the sphere to those of the cylinder circumscribed to it ; sphere equal in volume to a given cone or cylinder ; spherical segments. Several propositions remain obscure because Archimedes, addressing the savants of his period, takes these for granted. It was to remove these obscurities that EUTOCIUS wrote his Commentary, which is full of valuable historical information, *On the Measurement of the Circle*. A circle is equal to a right-angled triangle of which one of the sides of the right-angle is equal to the radius, the other to the circumference of the circle, i.e.

$$\pi R^2 = \frac{2\pi R \times R}{2}$$

Then the theorem which proves that the ratio of the circumference to the diameter lies between

$$3\tfrac{10}{70} \text{ and } 3\tfrac{10}{71}.$$

On Conoids and Spheroids. In this work, the curves of the second degree are defined by means of a plane section taken perpendicularly to the generatrix of a right cone. According as this cone is right-angled, obtuse-angled or acute-angled, a parabola, a hyperbola, or an ellipse is obtained. These curves, by revolution round their axes, generate what Archimedes calls a right-angled conoid (paraboloid of revolution), an obtuse-angled conoid (hyperboloid of revolution) and elongated or flattened spheroids (ellipsoids of revolution) (Fig. 7).

Amongst the results found by Archimedes, the following may be mentioned : The segment of the paraboloid of revolution is equal to one and a half times the cone having the same base and axis as this segment. Two segments cut off from a paraboloid of revolution

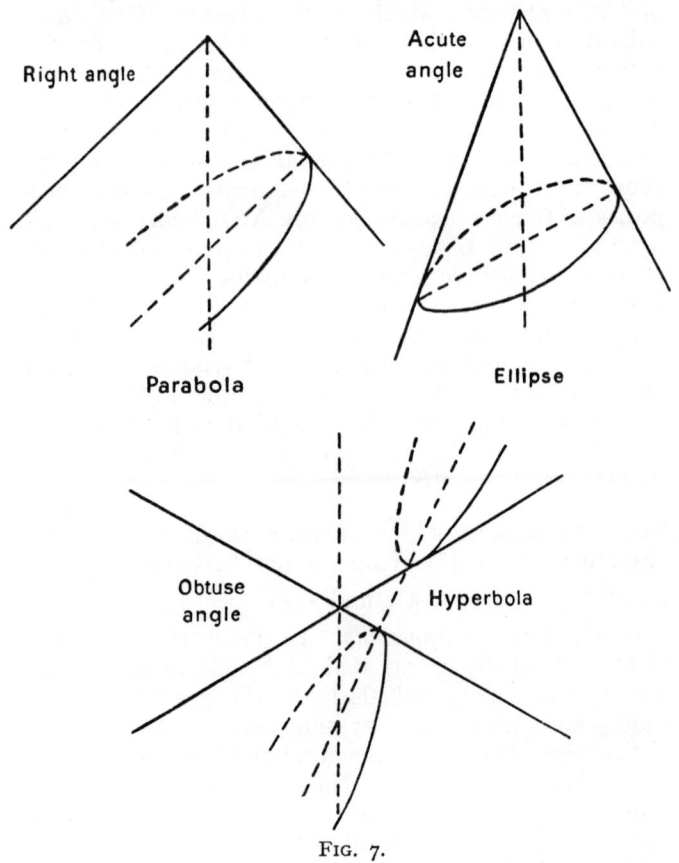

FIG. 7.

by any planes are to each other as the squares of their axes. To prove these demonstrations Archimedes uses the method of exhaustion, which consists in limiting the quantity sought between two known

THE ALEXANDRIAN PERIOD 75

quantities whose difference may be less than any given quantity.

The work *On Spirals* contains the study of the curve to which Archimedes has given his name, and which is described by a radius vector r increasing uniformly with the vectorial angle θ : $r = c\theta$, c being a constant.

The writing entitled *On the Equilibrium of Planes* or *The Centres of Gravity of Planes* is composed of two books. The first one begins by establishing the theory of the equilibrium of the lever, then enunciates and demonstrates various theorems relating to the centres of gravity of the parallelogram, triangle, and rectilinear trapezium. The second treats of the determination of the centre of gravity of a parabolic segment. The *Arenarius* is one of the most valuable documents we possess on the astronomy and system of numeration of the Greeks; amongst other things it contains a description of the heliocentric system of Aristarchus of Samos (Heiberg edition, II, p. 244, 12). To calculate numbers of any desired magnitude, Archimedes makes use of progressions, one arithmetical, the other geometrical, the former being used to find any term of the latter. *On the Quadrature of the Parabola* estimates the area of the parabola, first by means of pure geometry (method of exhaustion), then by considerations of equilibrium (infinitesimal mechanical method). The treatise *On Floating Bodies* establishes the fundamental laws of hydrostatics; the state of equilibrium of a liquid; the position of equilibrium of a solid immersed in a liquid according to the ratio of its density to that of this liquid. According to a legend related by Vitruvius (Bk. IX, 215, 10) Archimedes discovered the laws of hydrostatics whilst in his bath, thinking of the crown adulterated by the goldsmith of King Hiero.

The treatise *On the Method relating to Mechanical Theorems* has been recently discovered on a palimpsest of Jerusalem. In it new examples of the use of infinitesimal mechanical integration are described and worked out.[1] The *Lemmas* is perhaps an apocryphal work. As to the celebrated Cattle-Problem, it was propounded by Archimedes in the form of an epigram of forty-seven lines. It relates to the calculation of the number of oxen in a herd, being given that they are penned in order according to a regular figure, and that the animals of different colours occur in proportions successively dependent on one another. The work of Archimedes is so profound and original that we heartily endorse the judgment of Leibnitz: "He who understands Archimedes and Apollonius finds less to admire in the inventions of the greatest modern scientists."

APOLLONIUS OF PERGA (260–200 B.C.) is the third great mathematician of this period. Pappus represents him as vain and always ready to depreciate the worth of other geometers (Pappus, Hultsch edition, p. 678). In reality we do not know much about him, except that he was surnamed Epsilon, probably because the hall in which he gave his lectures bore the number $\varepsilon = 5$. He taught for several years in Alexandria, then in the university of Pergamum which had just been founded; after which he returned to Alexandria, where he remained until his death.[2] Of his masterly work on *Conic Sections* we only possess the four first books in the original Greek, the next three have been preserved in an Arabic translation, but the eighth and last is entirely lost. These books are dedicated partly to Eudemus, partly to Attalus, who is supposed by some to be Attalus I, King of Pergamum. In these dedications, Apollonius specifies the relation of his own

---

[1] See the articles of Th. Reinach and P. Painlevé in the *Revue générale des Sciences pures et appliquées*, November 30 and December 15, 1907.
[2] 23 Rouse Ball, *History of Mathematics*, p. 81.

THE ALEXANDRIAN PERIOD 77

discoveries to those of his predecessors. He shows how, in the first four volumes of his work, he has generalized and extended the elements of the theory already known. The third book enunciates new propositions which make it possible to solve a problem imperfectly treated by Euclid ; the fourth rectifies the results of Conon relating to points of contact and intersection of conics. The rest of the work contains further developments of the properties of conics and their applications.[1] In fact, what is really new in the work of Apollonius is his definition of conic sections. Archimedes and Euclid defined these as the sections taken perpendicularly to the sides of right cones, i.e. cones whose axis is perpendicular to the circle of the base, but of which the angle at the apex may be a right, obtuse, or acute angle (Fig. 7). Apollonius shows that the parabola, hyperbola and ellipse can be obtained by sections taken on one and the same oblique cone having a circular base. If through the axis of this cone we take a plane perpendicular to the circle of the base, we obtain the triangle formed by the two sides of the cone and the diameter of the base. If we now draw a plane perpendicular to the plane of this triangle, the sides of this triangle will be cut at two points, which will be the vertices of the curve. A similar geometrical construction will enable us to find a ratio indicating whether this curve or conic section is an ellipse, hyperbola or parabola. The geometrical constructions of lines and surfaces thus play the same part as algebraical equations in analytical geometry. But Apollonius not only expounds general theories, he applies them to numerous and difficult problems, carefully studying their conditions of possibility. The collections of these problems were for a long time in use in the school of Alexandria ; afterwards they were lost, with the exception of those

[1] 15 Heiberg, *Naturwiss.*, p. 56.

preserved in an Arabic translation. In one dissertation, unfortunately also lost, Apollonius examines the *Foundations of Mathematics*, and the fragments which have come down to us witness to his desire to connect mathematical concepts with reality, to reduce the number of fundamental propositions, and to justify their scope in the *Elements* of Euclid. Probably the other works published by Apollonius likewise had the aim of taking up again and investigating questions already studied by Euclid and Archimedes. For example a short work on *Unclassified Incommensurables*, and another on *The Dodecahedron and Icosahedron* are clearly inspired by Euclid, whilst the investigations of the *Helicoidal Line*, the *Contracted Method of Calculation*, and *The Burning Mirrors* were suggested by Archimedes. A treatise on *Contacts* of which many attempts at reconstruction have been made, must also be mentioned. Finally must be noted an astronomical treatise on the positions and retrogradations of the planets, which reveals Apollonius as the author of the ingenious theory of epicycles.[1]

As mathematicians belonging to the Alexandrian period, we must mention NICOMEDES, the inventor of the conchoid ($r = a \sec \theta \pm d$), and DIOCLES, the inventor of the cissoid ($y^2(2a - x) = x^3$)—these curves being used to solve the trisection of the angle and the duplication of the cube; and also GEMINUS, who wrote a valuable history of mathematics.

Whilst mathematics were advancing, practical mechanics also made remarkable progress as more and more importance was attached to engines of war used in besieging and defending fortified towns. The honour of having created the technics of this practical mechanics belongs to CTESIBIUS, a contemporary of Archimedes, who lived at Alexandria about the middle of the third century B.C. He constructed heavy

[1] 15 Heiberg, *Naturwiss.*, p. 58.

THE ALEXANDRIAN PERIOD 79

cannon, which were partly operated by compressed air. His works were unfortunately lost, but we find their essential features in the *Mechanics* published by his immediate successor PHILO of Byzantium, several fragments of whose work have come down to us.[1] A general introduction prefaces this most important work ; then comes the description of catapults, which it has been possible to reconstruct in recent times by the aid of the drawings which accompany the description. The accuracy and long range of these engines were a revelation.[2] Reflections on the art of besieging follow, then an accurate account of the theory of the lever, further on a description of automata and mechanical apparatus intended for use in theatres or gardens, such as magic goblets, water-cans pouring different liquids as desired, fountains with animals drinking and birds singing, etc. Beside these there were other more useful apparatus, such as for washing automatically the steps of the temples. The mechanism of all these machines is based principally on the action of levers and compressed air.

Two centuries later HERO of Alexandria took up the work begun by Ctesibius. He probably lived about the end of the first century B.C., but the dates of his life, death and works are very uncertain.[3] Although Hero of Alexandria is more famous in history than Ctesibius, his work is far from being of equal originality and accuracy.

From a mathematical point of view it consists of : (1) An elementary geometry, with applications to

[1] A. de Rochas, *La science des philosophes et l'art des thaumaturges*, Dorbon, Paris, p. 59.
[2] 10 Diels, *Antike*, p. 92.
[3] J. L. Heiberg and P. Tannery place Hero in the second century after Christ, but the majority of historians decide in favour of the first century before the Christian Era (17 Loria, *Sc. esatte*, p. 583, and W. Schmidt in his introduction to the works of Hero).

the determination of the areas of fields having a given shape ;

(2) Propositions on the method of calculating the volumes of certain solids, with applications to buildings used as theatres, baths, banqueting-halls, etc.

(3) A rule for finding the height of inaccessible objects.

(4) A table of weights and measures.

Amongst his writings on mathematics, must be mentioned, besides the *Definitions* and a *Commentary on the Elements of Euclid*, a recently discovered work on *Measurements*, in which the rules and formulæ for estimating the most important volumes and surfaces are enunciated together with theoretical proofs. The main part is borrowed from Euclid and Archimedes ; even the formula which gives the surface of a triangle in terms of its three sides a, b, c—i.e. $S = \sqrt{p(p-a)(p-b)(p-c)}$ (where $p =$ the semi-perimeter)—is not an original invention, for it was probably used by the Egyptian land-surveyors, and it is only the demonstration which can be attributed to Hero. He also attempted to perfect the levelling instrument hitherto used in surveying. These improvements are carefully described and theoretically correct, but they reveal the great practical ignorance of their author. The work entitled *The Construction of Vaults* was also probably written with a practical aim in view, and at any rate had the honour of being studied and commented upon by one of the architects of St. Sophia, Isidore of Miletus. Inspired by previous works, Hero has been able to give very exact information on *The Construction of Catapults* ; on the other hand, some of his writings, which are similar in conception to those of Archimedes and Philo, display great defects, especially the *Pneumatics*, in which the theory of the pressure of the air is applied to various apparatus. These are for the most part borrowed from Philo, and their descrip-

THE ALEXANDRIAN PERIOD

tion, containing some new additions, reveals an imitator who is unfamiliar with experiments and technique. The instructions given for the construction of a troupe of performing automata on a larger scale than that of Philo suffers from the same defect: the author, for example, forgets to describe the motive power which puts the whole in motion.[1] These latter writings were, however, much appreciated both by the Arabs and the savants of the Renaissance; they gave rise to the construction of many garden fountains with figures moving automatically, which excited the admiration of visitors. The old clock of Strasbourg with its moving figures is a direct descendant of the Automata of Hero. The *Mechanics*, of which we only possess the Arabic version, is less defective: it explains, in accord with Archimedes, the principles of statics and the parallelogram of forces, and describes the use of the toothed wheel, the lever, the tackle, the wedge, and the screw. Hero has also devoted a work to the study of the crane, and the problem of Archimedes: to move a given weight with a given force. Despite its defects, his work remains one of our chief authorities on the history of Greek mechanics.

2. GEOGRAPHY AND ASTRONOMY

The interest in geography awakened by the conquests of Alexander the Great, far from declining, continued and developed thanks to the fostering care of the Seleucids and the Ptolemies. The progress of mathematics, also, had a favourable influence on the development of this science, which, from the purely descriptive stage, grew more and more systematic and accurate.

ERATOSTHENES of Cyrene (275–194 B.C.), the learned librarian of Alexandria, must be regarded as the creator of geography as a science. His history of geography

[1] 15 Heiberg, *Naturwiss.*, p. 79 *et seq.*

since the Homeric age displays true historical perception, especially in comparison with the fantastic descriptions given at the same period by certain commentators on Homer. After having calculated mathematically the habitable regions (78,000 stadia by 38,000), Eratosthenes divides them by parallels to the equator and meridians into unequal rectangles. The parallel which passes through Gibraltar and Rhodes is in the middle and separates the northern parallels (Byzantium, Borysthenes or Dnieper, Thule) from the southern (Alexandria, Syene, and Meroe). The extreme meridians are formed by the Pillars of Hercules (Gibraltar) and the Ganges.[1] Eratosthenes also measured the length of the circumference of the earth by a method as ingenious as accurate. He observed that at Alexandria at noon, at the time of the summer solstice, the distance of the sun from the zenith is one-fiftieth of the circumference of the heavens, whilst at Syene at the same moment the sun is at the zenith, since it lights up perpendicularly the bottom of the wells. These two towns, situated on the same meridian, are 5,000 stadia apart. Therefore, by multiplying 5,000 by 50, the required measurement is found, namely, 250,000 stadia, which is equal to about 44,000,000 metres, the stadium being equal to 177·4 metres (Cleomedes, *de Motu circulari*, p. 96, 21). (Fig. 8.)

Although the method used is correct, the result obtained is not accurate. Firstly, Syene and Alexandria are not on the same meridian: between these two towns there is a difference of longitude of 3°.[2] Further,

---

[1] G. Lespagnol, *Géographie générale*, Delagrave, Paris, p. 83. For the authenticity and interpretation of the fragments of Erastosthenes, see A. Thalamas, *Etude bibliographique de la géographie d'Eratosthène*, Rivière, Paris, 1921 ; *La géographie d'Eratosthène*, Rivière, Paris, 1921.

[2] "Astronomie," *Kultur der Gegenwart*, Teubner, Leipzig, 1921, p. 187.

THE ALEXANDRIAN PERIOD 83

the length of 5,000 stadia, calculated by the day's march accomplished by caravans, is necessarily only approximate. The measurement found by Eratosthenes is nevertheless an interesting datum. Had Newton been acquainted with it, he would have been able to verify his hypothesis of gravitation, without being obliged to shelve it for years [1] until Picard succeeded in measuring the radius of the earth more exactly. In other realms of knowledge, Eratosthenes showed himself to be an erudite and remarkable savant, whom Archimedes held in high esteem, and with whom he wished to collaborate in his own researches. We do not know very much of his work, except the

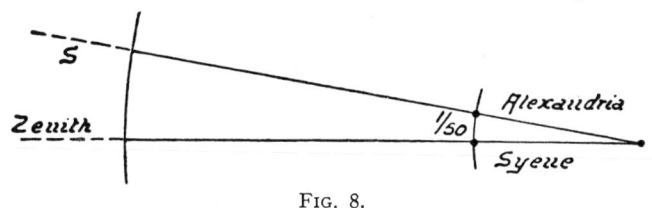

FIG. 8.

Sieve, which bears his name, which is a method of finding the sequence of prime numbers. He also invented, for finding the value of two mean proportionals, an ingenious mesolabe, which he placed in a temple of Alexandria with a dedication in honour of Ptolemy II. Finally, he devised the Calendar afterwards known as the Julian Calendar.

Astronomy, like geography, developed in a remarkable manner during this period, owing to the combined progress of mathematics, mechanics and technique. The surveying instruments with their screws and toothed wheels were of great assistance to astronomers, for instance the apparatus invented by Archimedes for measuring the diameter of the sun.

[1] 23 Rouse Ball, *History of Mathematics*, II, p. 16.

As a result, the observatory of Alexandria was able to undertake a systematically planned series of measurements to check the figures furnished by Chaldean astronomy. The customary divisions of the day and night being too inaccurate, the Babylonian division into exact hours, already known by Herodotus (II, 109), was adopted, and this, coming into current use, was afterwards accepted by the Romans. The sexagesimal division of the circle (degrees, minutes, seconds) was also borrowed from the Babylonians; but otherwise the Egyptian use of fractions having numerators always equal to 1 was preserved. The foundations of trigonometry were also laid. This is proved by a writing of Aristarchus of Samos (310–250 B.C.) in which, following the example of Eudoxus, he attempts to determine the magnitude of the sun and moon and their distance from the earth. The results obtained are satisfactory for the moon but not for the sun. In this writing Aristarchus keeps to the geocentric hypothesis, although, as we have seen, he elsewhere maintains the heliocentric hypothesis taken up by Copernicus many centuries later. The way for this hypothesis had already been prepared by the Pythagoreans and by the opinions held by certain groups of Athenian philosophers. It is also possible that Aristarchus was encouraged in his views by the influence of his master, the physician Strato. In spite of its simplicity, the heliocentric hypothesis was opposed for physical and religious reasons; for example, the Stoic Cleanthes considered it a blasphemy. Its only defender was SELEUCUS of Seleucis (about 150 B.C.), who gave at the same time a correct explanation of the ebb and flow of the sea, showing by observations the dependence of these phenomena on the position of the moon. He also affirmed, with Heraclides of Pontus, the infinity of the universe.[1]

[1] 15 Heiberg, *Naturwiss.*, p. 62.

THE ALEXANDRIAN PERIOD

Conon and Dositheus, the friends of Archimedes, were especially notable observers. Conon, in particular, discovered a group of stars which he called " Berenice's Hair " in honour of the wife of Ptolemy Euergetes.

Taking as a basis the celestial map of Eudoxus, Aratus of Soli wrote a descriptive poem on the starry heavens, which, although possessing no great literary qualities, made an enormous sensation. It had several Roman commentators, amongst them Cicero, and, with its illustrations of antique figures, enjoyed great fame in the Middle Ages.

However, the greatest astronomer of antiquity was incontestably Hipparchus, who was born at Nicaea in Bithynia and spent the greater part of his life at Rhodes. One of his observations on the star η Canis Majoris enables us (as Delambre has shown) to fix the date of his work at about the year 120 B.C. His scientific activity was prodigious. In his youth, he composed a *Commentary on the Phenomena of Aratus and Eudoxus*, which is unfortunately the only one of his writings now extant. He constructed several instruments, amongst others a dioptra for measuring the apparent diameter of the sun by a much simpler method than that of Archimedes. His apparatus is composed of a graduated scale on one end of which is a sight and on which slides a cursor. To take an angular measurement the cursor is moved until the eye looking through the sight sees it cover the magnitude to be measured, such as, for example, the diameter of the sun. This instrument with few modifications became that known as Jacob's staff, or cross-staff. Hipparchus also made use of two instruments to which he gave the name of astrolabe. " The first, or spherical, astrolabe was composed of several metallic circles, some fixed, others movable. The first circle of all was the meridian; it was suspended from a fixed point, or better still, supported by a small pillar, to which it was fixed by its lowest point; another circle, movable about the axis of the earth, could always be brought to coincide with the ecliptic; a third circle turned round the poles of the ecliptic on two cylinders which were fixed thereto and marked the longitudes; finally, a fourth circle, placed inside the three others, carried two sights used for sighting the heavenly body, whose position it was desired to determine." The second, or planispherical, astrolabe was quite different from the spherical astrolabe, or armillary sphere; it was composed of a disc which could be suspended vertically or placed horizontally; it was used for taking the altitude of the stars and for solving triangles. " Thus the same name has been given to two things which have no resemblance and thereby regrettable confusions have arisen."[1] Hipparchus also invented trigonometry, but, in order to solve a triangle, he always supposes it to be inscribed in a circle; the sides of this triangle are then chords which are calculated as a function of the radius of the circle. This being so, Hipparchus calculated a table of chords and laid down the formulæ by which the problems of spherical astronomy can be solved. He is thus more truly than Aristarchus the creator of trigonometry.

Having seen a new star appear, he had the idea of making for posterity a catalogue of the positions of the stars and principal constellations. One can never, said Pliny, praise him enough for this undertaking, which would have made even a god shrink back (*Nat. Hist.*, I, p. 159, 10). Thanks to the accuracy of his observations, which he compared with those of his predecessors, Hipparchus proved that, if the latitudes of the stars have remained constant, their longitudes have all increased by the same amount. He concluded

[1] Doublet, *Histoire de l'Astronomie*, Doin, Paris, 1923, p. 105.

THE ALEXANDRIAN PERIOD 87

from this that the vernal equinox is displaced along the ecliptic, and thus he discovered the precession of the equinoxes.[1] He propounded the problem, which bears his name, concerning the irregular movement of the sun, and he solved it by means of an eccentric movement which he calculated. He also discussed the irregularities of the moon and attempted to determine its parallax, and he thus succeeded in accurately predicting eclipses, which justifies the admiration of Pliny (*Nat. Hist.*, I, p. 143, 14). As Bigourdan remarks, " With this extraordinary man there suddenly appears a perfected astronomy, far superior to that of the preceding age ; the theories of the sun and moon are formulated, and those of the planets outlined ; the great desideratum of ancient astronomy, the prediction of eclipses, is now a problem solved. For the first time, the positions of a great number of stars scattered in the sky were known, and by the discovery of the precession their co-ordinates for any period could be calculated."[2] Hipparchus considered that geography as a science must be based on precise astronomical data; and he severely reproached Eratosthenes for not having satisfied this condition. But taking into account the difficulties of the work, these reproaches are unjust. Moreover they had the effect of retarding the scientific development of geography, which from that time became merely descriptive and ethnographical until its mathematical and astronomical aspects were once more studied by STRABO. The latter, however, looked upon exact science as only an occasional help to geography, the main work being to describe the countries known and inhabited in the time of Augustus, and not to make a study of the dimensions of the earth. Strabo certainly acquitted himself marvellously of his self-appointed task, particularly as

[1] Doublet, *Histoire de l'Astronomie*, p. 106.
[2] Bigourdan, *Astronomie*, p. 279.

regards Italy. His writings abound in narrative and vivid descriptions, gathered in the course of his travels from Armenia as far as Sardinia, and from the Euxine to Ethiopia. This period was rich in geographical literature, of which we only possess a small portion, comprising some fragments of POLEMON; and a description, by an unknown author, of Thebes in Greece as a town with somewhat unsafe streets, but charming with its fruitful gardens and veiled women.

Although the mathematical and astronomical side of geography was not neglected by Strabo, it is POSIDONIUS (133–49 B.C.) to whom it is most indebted. Posidonius was a native of Syria, but settled at Rhodes, where his school was frequented by Cicero and Pompey. Although a Stoic, he was interested in mathematics and natural science. He wrote an important work on the Ocean and a *Commentary on the Timaeus* of Plato, in which he treats of the mystic arithmetic of the Pythagoreans. Besides this, he was a champion of divination and astrology, the constructor of a planetarium, and a student of meteorology and astronomical problems. Geminus has given us a sketch of these works, and in the second century CLEOMEDES made use of them in his summary of astronomy (*de Motu circulari*, p. 90, 22). It certainly cannot be denied that Posidonius made original researches in geography and ethnography, but his claim to fame chiefly rests on the fact that he popularized and brought the principal geographical and astronomical attainments of the Greek science of his period within the reach of the cultivated public of Rome. In doing this, he often passes over in silence interesting theories, which thus, for long centuries, fell again into oblivion, for example, the heliocentric hypothesis of Aristarchus, and the explanation of the tides by Seleucus.

3. MEDICINE AND THE NATURAL SCIENCES [1]

Although Ptolemy II was a lover of curious and rare animals, the natural sciences made scarcely any progress during his reign; they remained as Aristotle and Theophrastus had left them. The writings on these subjects had a practical aim; the culture of fields and gardens, the raising of cattle. Certainly the poet CALLIMACHUS compiled a catalogue of birds, and the grammarian ARISTOPHANES of Byzantium wrote a history of animals, but these writers too often indulge in wonders and fables.

Medicine on the contrary, made real progress, largely due to the practice of dissection, which, forbidden in Greece, was practised in Egypt, favoured by the custom of embalming the bodies of the dead. It appears that the Ptolemies even authorized the physicians to make use of the living bodies of criminals condemned to death (Celsus: *de Medecina*, p. 4). Under these conditions an anatomy rapidly arose, founded on exact observation, and discovery followed discovery.

HEROPHILUS of Chalcedon is justly regarded as the creator of human anatomy as well as being the founder of the medical school of Alexandria. A disciple of Praxagoras (of the school of Cos), he avoided all dogmatism and made observation and experience the sole basis of his work. He discovered the nervous system and was the first to explain its nature and function; he also dissected the eye and the liver. In practical medicine he brought to light the importance of the pulse in diagnosis. In some respects, ERASISTRATUS of Ceos, the physician of Seleucus, was antagonistic to Herophilus. For example, he opposed the Hippocratic doctrine of the humours, and disapproved of the practice of bleeding, so much favoured in ancient

---
[1] 15 Heiberg, *Naturwiss.*, pp. 44, 46.

medicine. As an anatomist, he is remarkable: he distinguished between the nerves of sensation and those of motion, a distinction which had never before been made; he accurately described the heart, and recognized the importance of the brain, taking note of its convolutions. But he still believed with Praxagoras that the arteries contained air and not blood. If in wounds the blood spurted from the arteries, it was because there existed canals of communication between the veins and the arteries, and the blood, being no longer compressed by the air, passed from the former into the latter, conformably to Strato's theory of nature's abhorrence of a vacuum. The disciples of Herophilus and Erasistratus soon fell into a dogmatism, which brought about a reaction. A school arose called the Empiric, which confined itself to purely descriptive work and prohibited the inquiry into the general causes of things. At Rome medicine for a long time was in disfavour. CATO the Elder exhorted his son to distrust the poisonous potions of the Greeks; he recommended savoy cabbage as a remedy for all ills, and healed fractured limbs by magic words. But with the progress of civilization the need for physicians made itself felt. So that when ASCLEPIADES settled at Rome in the first century B.C. he met with immediate success. A native of Asia Minor, he was at first a rhetorician, but attained such distinction as a physician that he refused the offers of King Mithridates. He protested against the abuse of drugs and purgatives, he exalted the importance of hygiene and recommended cures by water, massage and exercise. In this way, without possessing very profound medical knowledge, he exercised a happy influence. Theoretically he adopted the humoral pathology of Hippocrates and completed it by Epicurean atomism. In fact Hippocrates remained the indisputable authority and his writings had many commentators, amongst them, APOLLONIUS

THE ALEXANDRIAN PERIOD 91

of Citium (50 B.C.). As the physicians were also pharmaceutists, botany benefited by their study of plants, for example CRATEVAS, the physician of Mithridates, wrote an excellent book on plants with illustrations and notes on pharmacy; and the poet NICANDER of Colophon composed a work on poisons and their antidotes, which in spite of its dulness found readers and commentators.

CHAPTER III

THE GRECO-ROMAN PERIOD

(From the Christian Era to the Sixth Century A.D.)

THE Roman Empire once established, Greek science was able to spread throughout the civilized world; it remained, however, foreign to the Western mind, while in the East it made some progress or remained stationary, before falling into decadence.

1. THE ROMANS AND SCIENCE

The Romans, owing to their essentially practical and political turn of mind, had little appreciation of pure science. They even despised it, and Cicero praises them because, thanks to the gods, they were not like the Greeks, and knew how to limit the study of mathematics to utilitarian purposes (*Tusculanae*, I, 2). The mathematical rudiments of which the Roman surveyors had need were borrowed from Greek writings in such a way as to enable them to be used in practice without the aid of theoretical knowledge. When need arose, specialists were called from Alexandria and shown the measurements to be made. It must have been in this way that Agrippa carried out the cadastral survey of the empire.[1] The fragments which appear in the mathematical compendiums are very poor. MARTIANUS CAPELLA (about 400 A.D.) published a work of bad taste, entitled *The Marriage of Mercury and Philosophy*, which was held in high

[1] 15 Heiberg, *Naturwiss.*, p. 73 *et seq.*

THE GRECO-ROMAN PERIOD 93

repute in the Middle Ages. In this work, he displays an utter incomprehension of mathematics by translating the first definition of Euclid, "the point is that which has no parts," by "the point is that of which the part is nothing." The works of Boetius, which, in the Middle Ages, were the basis of the teaching of geometry, arithmetic and music, have more value.

ANICIUS MANLIUS SEVERINUS BOETIUS (480–525 A.D.) belonged to one of the most illustrious families of Rome. At first a student, he afterwards unwillingly took part in the political life of his country and was notable for his charity and moral integrity. When elected consul, he tried to reform the coinage, but in so doing aroused hatred and envy, and being condemned by a tribunal was put to death, to the great regret of Theodoric. As a writer, he is well known by his *De consolatione philosophiae*. As to his book on *Arithmetic*, it is a rather crude copy of that of Nicomachus. In another work he gives without demonstrations the contents of the four first books of the *Elements* of Euclid, as well as some methods of surveying, drawn from various authors. This work is so little in agreement with what we know of Boetius that P. Tannery considers it a forgery, and Cantor supposes it to have been completely distorted by unskilful copyists. Such as it is, it contains a curious passage, which seems to describe a system of numeration based on the rule of position, the zeros being represented by empty places.[1]

If the Romans were antagonistic to pure science, they were, on the other hand, much addicted to superstitions. NIGRIDIUS FIGULUS by introducing astrology into Latin literature gained great fame amongst the cultivated classes. It was the same with the manual of astrology written with zeal and conviction by FIRMICUS MATERNUS. The short work of CENSORINUS

[1] 9 Cantor, *Geschichte*, I, p. 533.—6 Boyer, *Histoire des mathématiques*, p. 64.

on *The Day of Birth*, and the intelligent views of astronomy and physics, which Seneca, inspired by Posidonius, gives in a popular form in his *Naturales Quaestiones*, must be pointed out as worthy of interest. Amongst other subjects, Seneca devoted a long study to comets, to demonstrate that they must be likened to planets and consequently possess a periodic movement. The work of VITRUVIUS *On Architecture* is quite crude; the extracts from Greek authors on mechanics and technique are expounded so foolishly and in such obscure language that it would seem that the author, in spite of his pretensions, could not really have been an architect to Augustus.[1] The natural sciences are well represented by *The Natural History* of PLINY THE ELDER, whose death in 79 A.D. was caused by his desire to observe the eruption of Vesuvius from a near point of view. This vast compilation is a mass of observations collected with astonishing and often uncritical zeal and drawn from the most diverse writers; it brings before the reader a comprehensive survey of geography, anthropology, zoology, botany, medicine, mineralogy, and art. Perhaps the finest product of Roman scientific literature was the text-book of CORNELIUS CELSUS *On Medicine*. It formed part of an encyclopædia which has been lost, and although not written by a specialist, it makes intelligent use of Greek authorities and has preserved many an interesting detail, for instance, of Alexandrian surgery. Apart from the work of Celsus, there were only books of prescriptions. However, during the decline of antiquity many excellent translations of Greek authors appeared, such as the translation of the therapeutics of Soranus by CAELIUS AURELIANUS in the fifth century A.D. Works of this kind continued to appear until well into the Middle Ages, and even in the darkest periods the Greeks were acknowledged as the masters of medicine.

[1] 15 Heiberg, *Naturwiss.*, p. 75.

THE GRECO-ROMAN PERIOD 95

Geographical and ethnographical studies were much in favour amongst the Romans. Sallustus and Caesar give interesting information, the former on Northern Africa, the latter on Gaul. Tacitus describes Great Britain and particularly Germany and Scandinavia. In the only mention he makes of astronomical subjects, he shows how little the cultured Romans knew, for he explains the light of the polar nights by the flatness of the outermost countries of the earth, thus forgetting what had been a commonplace of knowledge in Greece for several centuries, to wit, the rotundity of our globe (Agricola, ch. 12). It is evident that the Romans did not study geography for its own sake, though we must except POMPONIUS MELA (first century A.D.) who utilized in a small but excellent text-book the statistical material collected by Agrippa.

2. GREEK SCIENCE IN THE EAST

Thanks to the power of tradition, intellectual activity was maintained, in spite of unfavourable conditions, simultaneously in Greece, Egypt and Asia Minor.[1] As soon as the imperial power came into the hands of the Antonines, Greek literature and science revived in some degree. There was a return to the past, which was specially favourable to the latter studies. The scientists were kept in practice by studying the great works of their predecessors, and if they made no original discoveries, they produced interesting commentaries, or systematized the results already obtained. Astronomy was brilliantly represented by CLAUDIUS PTOLEMY (date of birth uncertain, death probably 168 B.C.). Belonging to the Peripatetic School of philosophy, Ptolemy defended the views of Aristotle on the nature of matter and on gravitation; he maintained, for example, that a bather does not feel any pressure of the water above him, and that a bladder

[1] 25 Tannery, *Science hellène*, p. 5.

full of air is lighter than an empty bladder. His *Optics*, of which the first book has been preserved to us in a laborious translation from Arabic into Latin, treats not only of perspective as Euclid had done, but also of the physical conditions of vision and of optical illusions, and here Ptolemy accepts the theory of Plato that visual perception is produced by the rays proceeding from the eye meeting those proceeding from the object. In his *Catoptrics* he studies mirrors, and by measurements seeks to establish the law of angles of incidence and reflection. He also made comparative experiments on refraction in water and glass, and ascertained the existence of an astronomical refraction, the distance from a star to the pole being smaller when the star is on the horizon than when it passes the meridian. The figures found are not always accurate, but the experiments and ideas remain none the less of prime importance. Another work, more important still, was the one which Ptolemy devoted to astronomy. It was soon used as a text-book in the schools of Alexandria, and in order to distinguish it from similar but much smaller works, it was given the title of " ἡμεγίστη," the greatest (book understood), which translated into Arabic became corrupted into *Almagest*.[1] The work is divided into 13 books. In the first, Ptolemy gives an exposition of plane and spherical trigonometry and a table of chords. The second book discusses the phenomena arising from the spherical shape of the earth, with the admission that the hypothesis, which he rejects, of the revolution of the earth round its axis, would greatly simplify the explanations. Books III–VI treat of the movements of the sun and moon and of eclipses, all explained by means of epicycles and eccentrics. Books VII–VIII contain the catalogue of Hipparchus, completed and enlarged. The last books enumerate the sidereal phenomena which occur every

[1] 15 Heiberg, *Naturwiss.*, p. 82.

THE GRECO-ROMAN PERIOD 97

year. A set of well-arranged astronomical tables enables the time and eclipses to be determined according to the seasons and the days. These tables, because of their convenience, remained long in use.

The work thus accomplished is worthy of admiration, although Ptolemy lays himself open to the reproach of not having passed on to us any accurate observations, perhaps even of having made fictitious observations to justify his hypotheses.[1]

The *Tetrabiblos* is a compendium of astrology, which was wrongly, for a long time, not attributed to Ptolemy, being considered unworthy of him. It gives a systematic outline of astrological questions and contains many interesting ideas on the psychology of nations; it is far superior to similar works of that period. Amongst these must be mentioned the dialogue *Hermippus*, in which an unknown author defines the position of Christianity in relation to astrology.

Finally, in a geographical work, Ptolemy solves, with much skill, the problem of the projection of a spherical surface on a plane.

In the realm of mathematics MENELAUS published, towards the end of the first century A.D., a writing entitled *On Spherics*, which contains an important theorem on the spherical triangle. NICOMACHUS of Gerasa (Syria) brought out at almost the same time (A.D. 150) an *Introduction to Arithmetic*, which was, as we have seen, translated into Latin by Boetius. This introduction, amongst other propositions, enunciates the following: the cubes of whole numbers are successively obtained by the addition of odd numbers in this manner:

$3 + 5 = 2^3$, $7 + 9 + 11 = 3^3$, $13 + 15 + 17 + 19 = 4^3$, $21 + 23 + 25 + 27 + 29 = 5^3$, etc.

As to THEON of Smyrna, he is chiefly known by an

[1] 2 Bigourdan, *Astronomie*, p. 295.

exposition of the mathematical, astronomical and musical knowledge necessary to the understanding of Plato.

PAPPUS, who lived at Alexandria towards the end of the third century A.D., was remarkable for other reasons. He wrote several works of which we only possess one. This is a systematic account, with explanatory comments, of the great geometrical problems studied in antiquity. Designed as an aid to the understanding of the theories of Euclid, Apollonius and Archimedes, it contains a quantity of historical information of the greatest interest, the accuracy of which has often been verified. It is, besides, more than a mere compilation; in it we already find an enunciation of the theorem of Guldinus, the fundamental relation of the anharmonic ratio of four points, and the famous *problem of Pappus* on geometrical loci, the problem which was the starting-point of Descartes' researches on analytical geometry.

The dissertation of SERENUS of Antinopolis (Egypt) on the sections of the cone and cylinder do not contain anything very new; his proposition on transversals is of greater interest. However, it was DIOPHANTUS in particular, who, between the third and fourth centuries A.D., directed mathematics into a new path. His writings soon fell into oblivion, and it was not until the year 1460 that they became known to the scientific world through Regiomontanus. They contrast so much with the works of other geometers that some critics have found in them traces of Hindoo influence. Others, more enlightened, have recognized in them the contents, in a new form, of the geometrical algebra which had from the beginning been used by Greek mathematicians. It is scarcely credible besides that one man alone could have collected so many problems and solved so many equations. Diophantus had the great merit of creating a language and appropriate symbols: in doing this he has not altogether broken

THE GRECO-ROMAN PERIOD 99

with geometrical tradition, he still calls a square the product of two numbers; his method, on the contrary, is purely arithmetical. The problems are treated with much elegance, but point by point, without bringing in any general formulæ. The result is that Diophantus rejects as impossible the negative or irrational roots of an equation, and that, where two positive roots are possible he only keeps one. The problems set are very varied and lead to equations of the first, second, and sometimes third degree with one or more variables. One of these problems relates to the price of wine, and it is by the data of this problem that P. Tannery has fixed the period in which Diophantus lived.[1]

An interesting fact to be noted in the history of the mathematics of this period is the lively interest taken in them by the Neo-platonic school of philosophy. PORPHYRY and IAMBLICHUS devoted several writings to arithmetical questions, and PROCLUS in the fifth century A.D. wrote an interesting commentary on the works of Plato and the first book of Euclid.

Amongst other commentators of the period we must point out SIMPLICIUS, who, in 529 A.D., after the closing of the university of Athens by Justinian, fled into Persia, and whose commentaries on Aristotle are invaluable; also EUTOCIUS of Ascalon, to whom we owe an edition of the *Conic Sections* of Apollonius, and of some writings of Archimedes with explanatory notes. His work was rescued from oblivion by Isidore of Miletus, the architect of St. Sophia.

It was likewise to such commentaries that the later representatives of the mathematical school of Alexandria devoted their energies. THEON, about the year 370 A.D., edited the *Elements of Euclid* and the short course of astronomy which had been extracted from the Almagest for the purpose of teaching. His daughter HYPATIA, who fell a victim to the fanaticism

[1] 28 Tannery, *Mémoires scientifiques*, p. 70.

of Christian monks, commented on Diophantus and Apollonius.

If the exact sciences made but little progress, it was not the same with medicine.[1] A disciple of Asclepiades, THEMISON of Laodicea, founded the methodic school, who considered that all maladies arose from the general state of the body, a theory which might, however, lead to regrettable negligence of special symptoms. SORANUS of Ephesus was the most distinguished representative of this school in the second century. His literary output was very abundant and embraced all the subjects of medical interest, as well as the history of this science; unfortunately we only possess fragments of it, but these are sufficient to justify their author's reputation as a gynæcologist. Soranus treats not only of the child to be born and of the birth, but gives wise advice on the first cares to be lavished after the accouchement, on the choice of a wet-nurse, and on the treatment of abnormal and sickly infants. During the accouchement the mother must not be lying on a bed, but placed in a chair, specially constructed for this purpose. As to abortion, it must only be practised in an exceptional manner, and only in cases where the woman is unable to bring her child into the world without endangering her life. The newly-born babe must be nursed by its mother if possible. In any case, the meals must be regular, and the breast must not be given to quiet a child because it cries; for its cries, provided they do not last too long, are excellent exercise for the lungs. After a year and a half or two years the baby must be weaned, preferably in the spring.

In opposition to the Methodic school, there arose the Pneumatic school founded by ATHENAEUS (of Asia Minor), which connected its principles with the Stoic

[1] 15 Heiberg, *Naturwiss.*, p. 89 et seq.

philosophy. The spirit or pneuma (πνευμα), which is innate in every man, regulates health and disease. ARCHIGENES of Syria, about the year 100 A.D., somewhat modified this theory. His writings are lost, but we can reconstruct them partly by the quotations of Galen and partly by a compilation of ARETAEUS of Cappadocia, who borrowed from Archigenes the best part of its contents. It contains faithful and penetrating observations of nature, and a remarkable description of elephantiasis, a disease which was still unknown in the West. In therapeutics, Archigenes favoured regimen; he studied the effects of wine and mineral waters, and recommended cold water baths and sun baths.

Apart from some minor works of RUFUS of Ephesus, none of the medical literature of the first century A.D. is extant. This lack is due to CLAUDIUS GALEN, who played the same part in Greek medicine as Ptolemy in astronomy, that is, in his works, he absorbed and rendered useless those of his predecessors.[1] He was born at Pergamum in 129 and died at Rome in 200 A.D., received a careful and extensive education, and in the midst of a busy life, found time to write more than 150 medical works, of which about 60 are extant. This enormous production inevitably contains repetitions and superficial pages, and it is stamped with childish vanity, but it possesses none the less real merit, independently of the part it has played in the history of medicine. Galen indeed was not a mere compiler and arm-chair philosopher; he was a practitioner and knew how to carry out successful researches; he raised the level of medicine at an epoch when the schools in repute proclaimed, in the name of empiricism, the futility of theoretical preparatory studies for this science, and when it was

[1] For the life and writings of Galen, see Croiset, *Histoire de la littérature grecque*, V, p. 715, Fontemoing, Paris, 1899.

necessary to go from Rome to Alexandria to learn anatomy from a human skeleton. After having studied at Smyrna, Corinth and Alexandria, Galen, at the age of 28, settled at Pergamum as physician to the athletes. After some years he decided to try his fortune in Rome, in which city he soon gained great renown. When attacked by his colleagues he defended himself by publishing some pamphlets of which the tone and matter is often coarse. When he was about to be presented to the Emperor Marcus Aurelius, he abruptly quitted Rome, fearing that a plague, which had just broken out in the East, would spread there. He returned after a short time, and displayed great activity for another thirty years. His physiological conceptions are based on the humoral theory of Hippocrates, an author with whom he was very familiar and whom he followed intelligently; his doctrine of the vital forces placed by Nature in the body to control it, had a great influence in later times. In therapeutics, Galen recommends cures of fresh air and of milk, also medicines of doubtful composition. Amongst these, he highly commends theriac, an antidote against poison, specially prepared for the emperor, which was composed of 70 ingredients, including the bodies of boiled vipers. With all this, however, he recognized the importance of anatomy, and in default of human bodies the dissection of which was forbidden, he operated on animals, more especially monkeys.[1] After him, medical literature produced nothing but compilations of which the most celebrated is, justly, that of ORIBASIUS, the physician of Julian the Apostate.

Among the natural sciences, botany continued to benefit from the progress made by medicine. DIOSCORIDES of Cilicia in the first century compiled a catalogue of useful plants (to the number of 600),

[1] 15 Heiberg, *Naturwiss.*, p. 94.

THE GRECO-ROMAN PERIOD

which was very popular in the Middle Ages. Zoology, on the contrary, came to a standstill. Already in the second century, an unknown author, surnamed Physiologus, had foreshadowed by his fabulous and mythical descriptions of animals the Bestiary literature, and his work had a great influence on the animal decorations of the Middle Ages. In this period we must also mention Alchemy, to which we shall have to return and of which ZOSIMUS, about the year 300 A.D., summarizes the knowledge, sometimes fantastic, sometimes useful, relating to the working of metals.

PART II. PRINCIPLES AND METHODS

IN glancing at the history of humanity, one fact immediately attracts attention. It is the supremacy over all the continents which Europe has been able to win and to keep until the present day. The cause of this supremacy has not been either numerical superiority or a more advanced social organization or even any particular religious and literary ideas. The Chinese, as is well known, were civilized long before the Europeans, and, long before them, were acquainted with the use of the compass and even of gunpowder. The Hindoos, on the other hand, have possessed from the remote past a religion and a literature whose attraction, even to Western minds, is far from being exhausted; and in Central America there existed a state of advanced civilization, which was annihilated by the Spanish conquest. As to numerical superiority, it is sufficient to recall the fact, that even at the present time, either India or China has a larger population than Europe. If the white race has triumphed over other races, it is because it possessed weapons infinitely more formidable than those of its adversaries, and that for commercial transactions it had at its disposal manufactured products far superior to those of other nations. Now, the manufacture of these weapons and products has only been rendered possible through the progressive development of the mathematical and physical sciences of which the Greek nation laid down the principles and established the solid foundations. So it may be said that if ancient Greece had not created and transmitted rational science to Europe, the latter would never have gained and kept its world-supremacy. Doubtless, long before the Greeks, men possessed scientific knowledge, instinctive and practical. Already, in the Stone Age, they knew how to use the lever to move heavy objects, and how to make spears and arrows. At a later period the Chaldean and Egyptian civilizations witness to a very remarkable technical knowledge ; but as we have seen, they did not succeed in creating rational science, that is, in giving a reasoned explanation of natural phenomena and technical processes.

In the presence of Nature, two types of explanation can be utilized : the one brings into play the rational mentality, the other belongs to what M. Lévy-Bruhl calls the pre-logical mentality, and which it would be preferable to call with M. Brunschvicg the pre-scientific mentality.[1]

The latter is common amongst primitive peoples ; it conceives of the links of causality between natural phenomena as a form of mystical participation, which is in a sense extra-spatial and extra-temporal.[2] An individual is devoured by a crocodile or a lion. If he dies in this manner, it is not, in the mind of the savage, because he has imprudently approached one of these ferocious animals ; it is because a malevolent spirit has

[1] In fact, in the reasoning of the savage, the use of the principle of contradiction is by no means abolished as M. Lévy-Bruhl seems to imagine. Only it is exercised on another plane. To primitive man contradictions manifested themselves in the realm of the mystical, not in that of sensible experience. See our article, " Le problème de verité," in the *Revue de théologie et philosophie*, Lausanne, Dec. 1923. This is why we choose in preference to the appellation of M. Lévy-Bruhl that which M. Brunschvicg has adopted in his masterly work, *L'expérience et la causalité physique*, Alcan, Paris, 1922, p. 113.

[2] Lévy-Bruhl, *Mentalité primitive*, Alcan, Paris, 1922, pp. 55 and 516.

incited the crocodile or lion to devour him. These animals have not acted by themselves in obedience to their instincts, they are only an instrument used by the malevolent spirit. The latter could have chosen some other instrument, disease, for example, and in this case the individual destined to perish from its attack could have approached the lion or crocodile without danger. Here is another fact : Some one swallows poison and dies. To modern science, the poison, through the stomach, penetrates the blood, and corrupts it, or acts on the nervous system by causing an arrest of essential vital functions. There is here a whole chain of causes and effects produced from the moment when the poison is swallowed until that in which death supervenes. This succession of links is more or less rapid according to the case, and by the use of an antidote it may be checked. To the pre-scientific mind, things happen differently. It is an evil spirit, and he alone, who gives to the poison its hurtfulness ; by itself it has no power and without the spirit incarnate in it, would be harmless. Hence the custom of ordeals or judgments by poison, so common amongst savage tribes. Every accused person could vindicate himself by submitting to the test of poison ; if he vomited it, it was because he was innocent; if he died, it was because he was guilty. Thus, whilst the scientific mentality always seeks the cause of a sensible phenomenon in a combination of conditioning phenomena, also sensible, the pre-scientific mentality appeals to mystical and occult forces invisible and imperceptible to the ordinary means of perception. These forces are the real causes of sensible phenomena ; they float around man, who cannot always locate them in time and space, or even distinguish them, for they are in a sense extra-spatial and extra-temporal. They seem to imply to the primitive mentality a supplementary dimension ignored by us, not a spatial dimension like a fourth dimension, but rather a dimension of experience as a whole. We see that, for the linking of secondary causes which our sciences explain by formulæ and laws, primitive man substitutes another type of connection, that of occult and mystical powers. It is these powers which render effective the connections which we perceive between sensible phenomena in the effects of poison, drought, etc. It is therefore to these that heed must be taken for the guidance and right direction of life. Consequently the links which the scientist carefully notes in the succession of phenomena, have, for the primitive mind, only a relative importance, since they can be used indifferently by the occult power, and their connection is not inevitable. It is only the purpose of the spirits acting on these phenomena which needs to be considered, and not the means they use for its realization. Certainly savage races are not lacking in technical skill, and the pottery, baskets and canoes which they have succeeded in constructing with their clumsy tools are admirable. But this technical skill may be merely the result of long practice, it does not necessarily imply a scientific and thoughtful mental activity. It may be compared " to the skill of a good billiard player, who, without knowing a word of geometry or mechanics, without need for reflection, has acquired a rapid and sure intuition of the movement to be made in a given position of the balls."[1] To sum up, there is a profound difference between the conceptions of the pre-scientific mind and those of the rational mind. To the former, the production of each phenomenon is linked to the benevolent or malevolent disposition of the occult powers. The man may make use of certain talismans and practices to ensure the regular and favourable course of phenomena. By ritual prayers and sacrifices, fixed according to circumstances, he may propitiate the spirits and hence the

[1] Lévy-Bruhl, work quoted, p. 518.

events. But on the one hand it is not always easy to discover the really efficacious rite, and on the other hand, the desired result always remains uncertain since it depends on the good will of the spirits. The scientific and rational mind proceeds otherwise. In its conception the relation which unites one sensible phenomenon to another, such as a cause to its effects, is constant. Hence, this relation once discovered, the phenomena and the resulting consequences can be made use of with certainty.

Strange as it may appear, it is much easier to interpret natural phenomena according to the pre-scientific mind than according to the rational mind.[1] The actions and reactions which take place in nature are so complex and so varied that research into causes and laws in the scientific sense is extraordinarily difficult and arduous. In fact, no people except the Greeks have attempted it. The Hindoos, for example, in spite of their very advanced civilization, have never in their reasoning gone beyond the stage of the pre-scientific mentality. The flux of sensations which creates in us the image of the perceptible world does not, according to them, obey constant and fixed laws; it cannot give birth to a science, properly so called. Ancient Greece has had the genius and audacity to conceive that the matter on which our mental activity is exercised is subject to determinate relations. It has formed the opinion that these relations could not exist without a community of nature between the terms of which they are constituted: the effect must have some resemblance to the cause which produces it. It is

[1] M. Jean Piaget has just published a book which is very suggestive on this point, *Le langage et la pensée chez l'enfant*. Delachaux and Nietslé, Neuchâtel, 1923. This book, original in its method and results, shows in particular how, in the child, scientific notions are gradually and with difficulty substituted for pre-scientific and egocentric ideas.

the same in what concerns the relation of law to consequence.

This being so, it is necessary, for the explanation of the relations between the lines and surfaces of which geometrical figures are constituted, to have recourse to geometrical and numerical reasonings; to account for the phenomena of the physical world, it is necessary to appeal to mechanical and physical reasonings, and, finally, it is by physiological reasonings that health and disease must be explained, and not by invisible powers outside the body.

By these entirely new ideas the Greeks revealed to the human mind for the first time the true foundations of the sciences which, from the time of the Renaissance, were to blossom and give to Europe her supremacy. It may be objected with truth that these foundations had been laid already by the Egyptians and Chaldeans. But, as we have already remarked, these peoples had simply imparted to the Greeks mathematical facts and empirical formulæ which they had been able to establish through centuries of experience; they had never conceived of the possibility of creating a science worthy of the name. Between the fragments of knowledge which they discovered and the scientific conceptions of the Greeks there is an abyss which we may fathom by the following example. The Egyptians knew and made use of the numerical properties of the squares constructed on the sides of a right-angled triangle. We do not know how they discovered these properties, but it is probable, as we have remarked before (p. 7), that it was in the following manner. On the sides of a right-angled triangle whose magnitudes are 5, 4, and 3, let squares be described. We can divide these squares into smaller squares all equal to 1^2 and easily prove the equality $25 = 16 + 9$. This demonstration is purely empirical. It is so intuitive that a child can easily understand it. As it simply states a mathematical

PRINCIPLES AND METHODS

fact, it does not rest on any group of axioms or propositions previously demonstrated. It is complete in itself, but it lacks generality, since the sides of the triangle must be whole numbers of a certain value. Let us take, on the other hand, the theorem which tradition attributes to Pythagoras. We see immediately

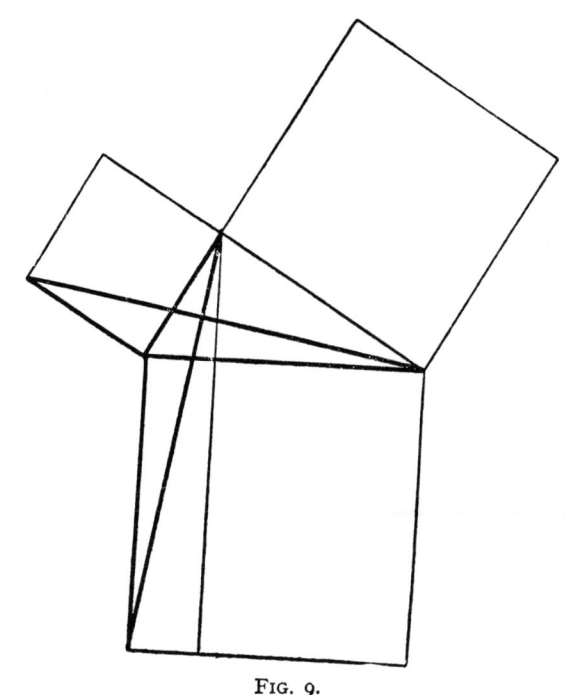

FIG. 9.

how different the demonstration is. The large square (Fig. 9) constructed on the hypotenuse, is divided into two rectangles ; the question being to demonstrate the equality of their areas with those of the squares constructed on the sides of the right angle. Auxiliary figures, viz. pairs of triangles, intervene ; this being so, it is necessary to prove first that the triangles of each pair are equal and then that one of them is equal in area to the half of one of the squares, etc. The demonstration in this form is quite general, independent of particular cases, but it supposes a whole series of propositions previously demonstrated and which are rigorously linked together; for example, all triangles which have the same base and the same height as a rectangle have equal areas, which are equivalent to half that of the rectangle.[1] To establish all these propositions, they must be based on the general properties of the straight line and the angle, in other words, on axioms and definitions. These axioms or definitions must be logical and in no way obscure to the mind, otherwise the deduction would remain doubtful and would lack exactitude.

Thus, the ideal which the Greeks have more and more conscientiously pursued is the following : to place at the basis of all science a number of principles which guarantee a strict logical reasoning, and then by their means to construct an edifice of consequences the value of which is assured by a rational deduction. Without insisting further it can be seen how much the Greek ideal of knowledge differed from that of primitive peoples or even of the peoples of the East.

[1] In this demonstration the investigation of the congruency plays a preponderant part as M. E. Meyerson rightly remarks: *De l'explication dans les sciences*, vol. I, p. 137 *et seq.*, Payot, Paris, 1921).

CHAPTER I

THE MATHEMATICAL SCIENCES

1. THE PURPOSE AND SCOPE OF GREEK MATHEMATICS

WHEN we consider the questions studied by the Greek mathematicians, we are at first astonished at their great diversity. Besides completed works, we find in the compendium of Diophantus the principles of a theory of numbers, in Apollonius the first idea of an analytical geometry, in Archimedes the clear conception of the infinitesimal calculus, and in Euclid the almost perfect application of a method of exposition which has remained the basis of more modern works.[1]

Important as they are, these discoveries only embrace a portion of the vast field of mathematics. The relations of numbers and figures constitute a world so extraordinarily complex, that much of it is still unexplored by modern science. And amongst all the aspects of this world of relations, the Greek scientists have been obliged to make a choice. What have been the reasons and circumstances which determined their choice? It is on this question that we must attempt to shed some light.

On the nature of the mathematical fact there is unanimous agreement. The Greek mathematician admits implicitly or explicitly that the science of number and space deals with ideal objects, changeless and incorruptible. Plato has powerfully expounded

[1] 4 Boutroux, *Idéal*, p. 31.

this manner of thinking, supporting it by metaphysical arguments. The mathematical sciences cannot be founded on the unstable and changeful phenomena of the sensible world; for instance, the aim of geometry is the knowledge of the eternal, and hence it attracts the soul towards truth, and makes it look upwards instead of downwards; arithmetic likewise has the virtue of elevating the soul by compelling it to reason about abstract numbers, without ever suffering its calculations to revolve about visible or tangible objects (*Rep.* 525 D). Thus there exists a world of notions or ideas which is complete in itself, and which has no need of support from the sensible world. These notions or ideas maintain between themselves immutable relations, the discovery of which is the province of the human mind.

On this point, all the Greek geometers, whether they accept or reject the Platonic idealism, are in accord. The figures about which we reason are not those perceived by our senses. There does not exist in reality any point which has no parts, any line without breadth, or surface without thickness. The material figures aid the imagination and thus are a help to the reasoning, but they are only an accessory aid. What constitutes the essential character of a geometrical figure, what causes it to be a mathematical entity, is the connection, defined once for all, of its component parts. Let us take, for example, the circle. Having once postulated the notions of straight line, distance, equal distance, we create, so to speak, the circle ideally, declaring with Euclid (Definition xv, *Elements*, I, p. 4) that a circle is a plane figure, bounded by one line, and such that from one interior point we can draw to this line straight lines all equal to one another. The circle thus created has no definite magnitude in the imagination, for it may represent a microscopic surface just as well as a region extended as far as desired into space. The definition

THE MATHEMATICAL SCIENCES 115

of a circle may therefore take a concrete form in sensible representations, but it is not exhausted by any of them, and it is not these representations which justify its existence, for they are never anything but an imperfect image. As will be seen, the definition sheds light upon the structure of mathematical principles and shows them to be distinct from the data furnished by sensible perception. This distinction impresses itself on the geometer apart from the metaphysical reasons, always debatable, by which it may be justified. What is certain, is that the principles of mathematics, thanks to their definition, can serve as a basis for strict reasoning, which can never be contradicted by any sensible experience. If we take at random two points on a circumference and if with these two points and the centre of the circle as vertex, we construct a triangle, we can affirm that this triangle is isosceles and has two equal angles. This affirmation is directly derived from the definitions which have been given of the isosceles triangle and of the circle. Thus to the Greeks belongs the great merit of having demonstrated that numerical expressions and geometrical figures possess peculiar properties of their own, judged by other criteria, and dependent on other methods of investigation than the phenomena of the sensible world. But this does not enable us to understand what has guided them in their choice of the innumerable problems presented by arithmetic and geometry. Doubtless it is very important to recognize the quality of the materials and the way to utilize them for the construction of a building, but it is also necessary to sort them according to the plan of the building. Now, the regular combinations of numbers or figures are unlimited in number. Analytical geometry has revealed to us several curves (the curve called by French mathematicians *la courbe du diable*, for example) of which the Greek scientists had not the slightest idea. Why did they stop at a certain property of numbers or a certain class of figures rather than at any other? The standard by which they made their choice of figures was the construction. This construction, as P. Boutroux points out, has nothing in common with the concrete measurements of surveyors. "It is a rational operation by which the theoretical *existence* of the figures on which the reasoning is based can be stated and proved. To attain this object, the most simple means evidently consist in constructing the figure, or rather in defining a theoretical process which would permit the construction to be made if it were possible to draw perfectly." [1] It is quite possible, however, to conceive of a figure being constructed or drawn by means of straight lines and circles, or even by considering the path traced by a point which moves on a plane or in space according to a given law (cycloid, spiral, etc.). Here a choice need not necessarily be made. The Greeks, after some hesitation, would only admit as legitimate constructions those which could be made by means of the straight line and the circle, or, in concrete terms, by means of the rule and compass. The objects of plane geometry are thus clearly defined. In dealing with spatial geometry, however, a difficulty at once arises. Solid bodies cannot be represented by a plane drawing without using descriptive geometry. The Greek geometers did not think of having recourse to this expedient, and did not at first know how to get over this difficulty, for which Plato reproaches them very severely (*Laws*, 528 B). They ended by admitting *a priori* the legitimacy of constructions, which correspond spatially to plane constructions made with rule and compass; the construction of a plane, a straight line or a circle in space, and also of round bodies such as the cylinder, cone, sphere, generated respectively by the revolution of a rectangle, triangle, and circle round a

[1] 4 Boutroux, *Idéal*, p. 38.

THE MATHEMATICAL SCIENCES 117

rectilinear axis.[1] At the same time conic sections took their rightful place in geometry, since they could be obtained, as we have seen, by the intersection of a cone and a plane suitably placed. Such curves as the quadratrix of Hippias, the conchoid of Nicomedes and the cissoid of Diocles remained rather on the margin of the pure and officially recognized science; they were considered too mechanical because instruments other than the rule and compass were needed for their construction.

Descartes rightly points out how arbitrary such a distinction is. I cannot understand, he says in effect, why the Ancients called these curves mechanical rather than geometrical. "For if we say that it might have been because it is necessary to use some instrument to describe them, it would be necessary to reject for the same reasons circles and straight lines, since these can only be described on paper by means of compass and rule, which may also be called instruments." [2] The argument of Descartes appears to be unanswerable. But then, whence came the self-imposed limitation of the Greek geometer? According to P. Boutroux there was no other reason for this but the desire to obtain a science which was simple and well arranged and consequently beautiful and harmonious. This reason does not seem absolutely decisive. Certainly the tracing of a straight line or a circumference is done by means of a very simple process; besides, the straight line and the circumference represent perfect and homogeneous mathematical facts, for two arcs of the same circumference can be superposed just as two sections of the straight line, but there the simplicity ends. As soon as the relation of the radius to the circumference is sought, the problem becomes obscure. Hence the fruitless attempt to effect the quadrature of

[1] 4 Boutroux, Idéal, p. 40.
[2] Geometry, Bk. II; edit. Adam and P. Tannery, VI, p. 388.

the circle, which Greek geometry made from its beginnings and never ceased to pursue.

The search after harmonious simplicity is not sufficient by itself to explain the direction of Greek mathematics. It seems to us that it is necessary to add on one hand the influence of the technical arts, and on the other the fear of clouding reason by bringing in mechanical means other than the rule and compass. The first point appears to be beyond question. As G. Sorel has repeatedly pointed out, it was certainly from the art of the engineer and the architect that Greek geometry borrowed its primary problems and, up to a certain point, its definitions. Thales was an engineer as well as a geometer; according to a tradition, which appears to be true in spite of the reservations of Herodotus (I, 75), he diverted by a canal the waters of the river Halys and rendered it fordable by the armies of Croesus. It must not be forgotten either that the father of Pythagoras at Samos was an engraver of seals. These possessed a magical value universally recognized, and the glyptography of Samos was famous for its productions.[1] Perhaps the invention of regular polyhedra ought to be attributed to the stone-cutters whose fumblings must have preceded the reasonings of geometers. G. Sorel believes likewise that a considerable part of the *Elements* of Euclid is derived from the art of building. He considers that the definition XXIII of parallels as straight lines produced to infinity and never meeting, is an interpolation, because it is not in keeping with the necessity for Greek geometry of avoiding the direct intervention of infinity. Euclid certainly ought to have defined the parallelism of two lines as a function of their equidistance. He was only translating into geometrical language the practice of architects, who for the construction of a wall use

[1] G. Sorel, *De l'utilité du pragmatisme*, Rivière, Paris, 1921, p. 198.

THE MATHEMATICAL SCIENCES 119

rectangular blocks carefully cut in such a way as to be able to interchange them in their superposition. Further, the obscure definition of a straight line given in the *Elements* (definition IV) takes on a new light if considered in connection with the art of the mason. The latter in order to verify the facing of a chiselled surface applies to it a stone rule coated with red oil. If the facing is perfect, the imprint made by the rule appears without any break; if not, there are gaps. Hence, the definition of the straight line " as a line lying equally between its points." However, it seems that Greek geometry, as it progressed, was able to free itself from the shackles laid on it by the age-long use of the rule and compass, and to conquer new and vaster realms by adopting figures constructed by other means.

If it has not accomplished this, it is doubtless because of the contempt in which tools fashioned and handled by slaves were held; [1] but it is probably also because the geometrical tracings obtained by these instruments raised problems insoluble by logic, for the following reasons: The instruments by which figures can be described mechanically may be divided into two groups: the first comprises the instruments whose arrangement remains exactly the same whilst the figure is described; for example, the legs of a pair of compasses keep the same length and the same opening, while one of them traces the circle. In the same way a triangle which generates a cone remains identical in area and length

---

[1] As M. E. Meyerson reminds us, Plato, speaking of the geometrical demonstrations into which mechanics enter, declares that this is to degrade geometry by making it pass, like a fugitive slave, from the study of things incorporal and intelligible to that of objects perceptible by the senses, and by using, besides reasoning, objects laboriously and slavishly fashioned by manual labour. *Bulletin de la Société française de philosophie*, Feb.–Mar., 1914, p. 101.

of sides, and it is this identity which logically guarantees the properties of the figure generated. The instruments which form the second group are, on the contrary, composed of one or more parts which change their respective positions whilst the figure is described. Consequently these parts do not occupy the same position at the beginning and the end of the operation. In tracing a quadratrix, for example, the radius of the circle moves angularly whilst the straight line which cuts it moves so as to remain constantly parallel to itself (p. 55). How can the point of intersection resulting from the combination of these two movements be logically defined? This intersection involves the indefinite divisibility of the radius and the straight line, and thus runs counter to the objections raised by Zeno of Elea. It would seem that it was a reason of this kind that consciously or unconsciously impelled the Greek geometers to admit only figures constructed by rule and compass, and the solids of revolution generated by these figures.

2. ARITHMETIC AND ALGEBRA

The Greek scientists took little interest in concrete applications of science, and they early distinguished between theoretical arithmetic and the art of calculating numerically concrete magnitudes. According to Plato's saying, we must reason about numbers as abstractions and not about numbers which are visible and tangible (*Rep.* 252 D). Hence " when we speak of Greek arithmetic, we understand the theory of the properties of numbers and exclude all that concerns calculation, namely, that which, since Plato at least, has been called logistic."[1] A scholium on the *Charmides*, translated by P. Tannery,[2] endeavours to define

[1] 25 Tannery, *Science hellène*, p. 369.
[2] 26 Tannery, *Géo. grecque*, pp. 48 and 49.

what must be understood by this science, as distinct from pure arithmetic. Inspired by this scholium, P. Boutroux justly points out that " Far from likening magnitudes to numbers, according to Greek tradition it was not permissible to consider as true numbers the numbers resulting from measurements of magnitudes, such as phialitic numbers, or relating to phials, melitic numbers, or relating to apples (or flocks). And this is why problems dealing with magnitudes were enunciated in concrete and not theoretical terms ; what for us is the ' solution of an equation of such or such type ' was formerly the solution of the problem of the oxen, the problem of the trees, the problem of the rabbits, etc." [1] Even in our own times schoolboys speak of the problem of the runners, the problem of the fountains, etc.

At first, however, the distinction between logistics and pure arithmetic was not clearly defined. It is certain that though Euclid surpassed the knowledge of the Pythagorean school, he left aside many of the questions studied by it.[2] The Pythagorean arithmetic was certainly more varied in its researches and, up to a certain point, in its conceptions, than the arithmetic of its successors. The fact is easily explained.

Although the Pythagoreans had the indisputable credit of laying the foundations of mathematical science in Greece, they were not able to free them from all metaphysical considerations. This fact is especially striking in regard to arithmetic, which was in a sense the corner stone of the Pythagorean philosophy, in whose eyes number and its properties constituted the basis of reality. In truth, sensible phenomena which are most diverse from a qualitative point of view, can show identical numerical relations. There is, for example, from the standpoint of the impression received,

[1] 3 Boutroux, *Analyse*, I, p. 121.
[2] 25 Tannery, *Science hellène*, p. 370.

a great difference between the shape of a right-angled triangle and that of a scalene triangle ; nevertheless, if the bases and heights of these triangles be equal, their areas will be expressed by exactly the same number. A regular hexagon and an equilateral triangle appear to us very different, but the hexagon can be decomposed into six equilateral triangles.

But it is not only motionless figures which can be measured, the movements of the stars are likewise subject to the law of number. And furthermore, musical sounds are heterogeneous as to quality with respect to each other, for a number of low notes cannot produce a high note and inversely ; but there exist numerical relations between the quality of sounds and the dimension of the objects producing them. Thus number is at the basis of everything. To the Pythagoreans it was not an abstract symbol, but a concrete reality,[1] occupying a determinate place in space, having clearly defined qualities and affinities, both moral and physical, something like the chemical atom. Under these conditions numbers are identified with space, they not only resemble it, but they create it. Thus, by a suitable analysis it is possible to find groups of numbers which correspond to certain spatial forms. According to the Pythagoreans the best analysis is that obtained by means of the gnomon or set-square. As defined by Hero of Alexandria (iv *Definitiones*, p. 44, 13) the gnomon is that which, being added to a number or figure, gives a whole similar to that to which it has been added.[2] This being so, let us suppose a set of gnomons (or set-squares) which fit into one another. If the first encloses one point, the second three points, etc., then it will be seen that the sum of the uneven numbers forms squares (Fig. 10). If the gnomons enclose even numbers, the result is no longer squares,

[1] 7 Brunschvicg, *Etapes*, p. 34.
[2] 20 Milhaud, *Phi. géo.*, p. 88.

THE MATHEMATICAL SCIENCES

but rectangles (Fig. 11). We notice also that the sum $1 + 2 + 3 + ... + n$ of n consecutive numbers beginning by one is a triangle (Fig. 12).

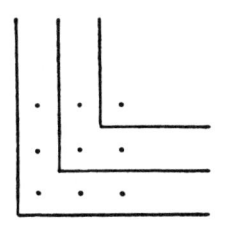

It is not only plane figures which thus correspond to sums of numbers arranged in series, it is also spatial figures. For example, by superposing the triangular numbers we obtain the pyramidal numbers 1, then $1 + 3 = 4$, then again $1 + 3 + 6 = 10$, etc., this being represented as in Figure 13.

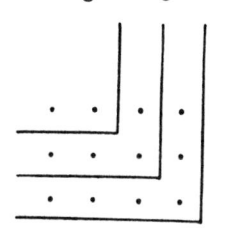

FIG. 10.

FIG. 11. 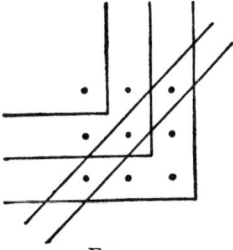   FIG. 12.

It was probably from these arithmetical-spatial conceptions there originated the classification of numbers into squared numbers (obtained by multiplying a number by itself), plane numbers (formed by two factors), and solid numbers such as the cube. Of this classification only the terms square and cube still remain.

Further, as numbers were not abstractions, but beings endowed with qualities and almost feelings, there were some which were perfect, that is, equal to the sum of their divisors (for example, $6 = 1 + 2 + 3$), and there were others which were "friendly," that is, such that each was equal to the sum of the divisors of the other.[1]

[1] 3 Boutroux, *Analyse*, I, p. 5.

124 SCIENCE IN GRECO-ROMAN ANTIQUITY

According to G. Milhaud it is possible to explain by arithmetic the table of metaphysical categories framed by the Pythagoreans.[1] This table sets forth, on the one hand, the ideas of " finite," " odd," " unity,"

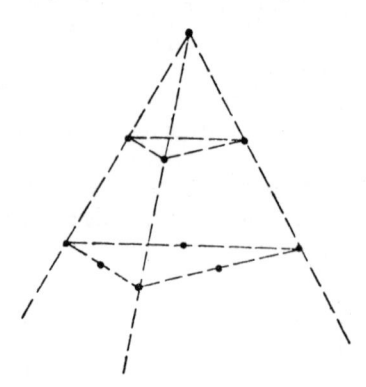

FIG. 13.

" square," etc., with, on the other hand, the opposite ideas of " infinite," " even," " plurality," " heterogeneous factors," etc. In order to understand these oppositions we must remember this: if we build up the odd numbers with the gnomon, we obtain a square, i.e., a finite and complete figure, whose sides have a ratio $\frac{n}{n}$ always identical and equal to unity. On the contrary, the construction of the even numbers by the gnomon gives a rectangle, a figure indefinite in this sense that its sides n and $n+1$ have a ratio changing with the value of n, namely: $\frac{2}{3}, \frac{3}{4}, \ldots \frac{n}{n+1}$

We know also that, in their arithmetic, the Pythagoreans went so far as to consider that even moral realities were formed of numbers.[2] Justice, for

[1] 20 Milhaud, *Phi. géo.*, p. 116 *et seq.*
[2] 22a Robin, *La pensée grecque*, p. 73.

THE MATHEMATICAL SCIENCES

instance, was identified with the number four, the square representing perfect equilibrium. Nevertheless, in spite of their metaphysical and mystical tendency, the Pythagorean researches led to several interesting discoveries. Besides the properties of certain series of numbers, they have defined different types of means :

1. The arithmetic mean such that $a + b = 2m$,

$$\frac{a-m}{m-b} = \frac{a}{a}$$

2. The geometric mean such that $m^2 = ab$,

$$\frac{a-m}{m-b} = \frac{a}{m}$$

3. The harmonic mean such that $\frac{2}{m} = \frac{1}{a} + \frac{1}{b}$,

$$\frac{a-m}{m-b} = \frac{a}{b}$$

But these proportions had no meaning for the Pythagoreans unless they were formed of whole numbers; they do not apply to any kinds whatever of magnitudes, commensurable or not, even when these are proportional. However, the advance made by spatial arithmetic through the Pythagorean school was checked on the one hand by the discovery of the irrational $\sqrt{2}$, and on the other by the criticism of Zeno. Besides, the mystical speculations on which this science appeared to rest became more and more repellent to the minds of scientists desirous of obtaining positive results. The consequence was that, amongst the Greeks, arithmetic made little or no progress.

Euclid, however, systematized in Books VII–IX of the *Elements* the results which had been obtained.[1] He represented numbers as lengths, and deduced their

---

[1] 23 Rouse Ball, *History of Mathematics*, I, p. 62.

properties from those of geometrical figures. He studied the theory of rational numbers, indicated the rules for finding the greatest common factor and the least common multiple ; he also studied fractions and geometrical progressions and demonstrated that the number of prime numbers is unlimited.

Is it to the system of numeration in use amongst the Greeks that their lack of progress in arithmetic should be attributed ? Certainly this system was not as practical as our own, but this was not an insurmountable barrier, as is shown by the *Arenarius* of Archimedes.

However this may be, arithmetical speculations were only revived in Greece by Diophantus and then in an algebraical form. The originality of Diophantus consists in the first place in having entitled his work ἀριθμητικά (Arithmetic), and then in treating of matters which are logistical. This innovation was more than a matter of words, it brought into abstract science that which had formerly been considered to belong to concrete science ; it announced a change in form and method. With one exception (*Opera* I, p. 385) the numbers of Diophantus are abstract and do not relate to oxen or rabbits ; the problems also are treated methodically, their solution is not merely enunciated without demonstration, as had been the case with the logisticians.

Although Diophantus had eclipsed all his predecessors, his aim was not understood in the way he desired. Nicomachus, in his treatise on arithmetic, still considers the numbers of Diophantus as concrete. The traditional distinction between arithmetic and calculation remained, although the deep abyss which separates them is henceforward filled up.[1]

As a matter of fact, the Arabs did not translate Diophantus until the tenth century, and it was only

[1] 26 Tannery, *Géo. grecque*, p. 52.

in the year 1575 that he became known to the Western world.[1]

3. THE IRRATIONAL $\sqrt{2}$. THE ARGUMENTS OF ZENO OF ELEA. PROPORTIONS AND THE METHOD OF EXHAUSTION. INTEGRAL CALCULUS.[2]

The arithmetical realism, naïvely proclaimed by the Pythagoreans, was checked by the discovery that in a square the diagonal and the side are incommensurable. If space be number or ratio of numbers, this discovery is disconcerting. The Pythagoreans doubtless did not pretend to estimate the number of points which compose a segment of a straight line, but they affirmed that this number exists, and that it is necessarily a whole number, since the point is indivisible. Between two straight lines A and B of unequal length, there must be the ratio A/B, in which A and B, representing a sum of points, are necessarily two whole numbers. This ratio leads in fact to a more simple ratio n/N, if a suitable unit of measurement be chosen to estimate the lengths A and B, since this now plays the part of common factor. Let us now suppose that the sides of a square each have 10 times the unknown number of points. According to the so-called theorem of Pythagoras, the square described on the diagonal will contain 200 times this number. The diagonal must therefore be equal to a whole number which, multiplied by itself, gives exactly 200. Now 14 is too small, for $14 \times 14 = 196$, and 15 is too great, since $15 \times 15 = 225$. Then let us take the side of a square equal to not 10 times but 100 times, to 1,000 times, to n times the number of points, etc. Whatever be the figure chosen, we shall never find for the diagonal a number which

---

[1] 23 Rouse Ball, *History of Mathematics*, I, p. 118.
[2] See our book, *Logique et Mathématiques*, Delachaux, Neuchâtel, 1900, and our article in the *Revue de Metaphysique et Morale*, July 1911, " Infini et science grecque."

when squared will exactly equal 2×10^n. Of this fact the Pythagoreans were able to give the following demonstration. Let a be the diagonal and b the side of the square. These two numbers may be supposed to be prime to one another, for if they were not, they could always become so by the suppression of their common factors. From the equation $a^2 = 2b^2$ we must conclude that a^2 and consequently a is an even number. Since a and b are prime to one another, b can only be odd. But if a be even, we can postulate $a = 2a_1$ and the original relation becomes $4a_1^2 = 2b^2$ or $2a_1^2 = b^2$. In this case b is even, but then a and b are no longer prime to one another, which is contrary to the hypothesis. The side and the diagonal of a square are thus incommensurable.

Although disconcerted by this discovery, the Pythagoreans regarded it as an isolated instance; it did not cause them to modify their arithmetical-spatial conceptions, and they were not able to glimpse the relationship between the continuum and infinity. Zeno of Elea was the first to propound this problem with precision. According to a generally accepted opinion, he desired, in discussing this question, to prove first of all the impossibility of motion, and, indirectly, to deny the plurality of Being. But, as we have seen, a passage of Plato (*Parmenides*, 128 C) shows that Zeno simply sought to oppose the idea of plurality as affirmed by the Pythagoreans. The testimony of Plato is the more convincing since the argument of Zeno has no significance if it denies the fact of motion, but is, on the contrary, decisive in showing that motion is incompatible with the hypothesis of plurality. Of this argument briefly summed up by Aristotle (*Phys.* 239 b 9) we only possess the parts which deal with continuity in its relations with infinity.

According to Zeno it must be admitted that either the division of space, time and motion can be continued

THE MATHEMATICAL SCIENCES 129

indefinitely or else that it has a limit. Let us suppose in the first place that the division be indefinite. In this case a moving body cannot traverse the length AB because before reaching the point B, it must traverse the length $\frac{AB}{2}$ and, before that, $\frac{AB}{4}$, $\frac{AB}{8}$, etc. The dichotomous division of AB being infinite, one cannot see how the displacement of the moving body can be produced. There is the same difficulty if we consider the relation between two objects in motion. Achilles runs ten times faster than a tortoise, but if he gives it a start of ten yards he will not be able to overtake it. The space he would have to traverse in order to do this is represented by the sum of the following stages, the length of which certainly diminishes but never becomes zero :

$$10 + 1 + \frac{1}{10} + \frac{1}{10^2} + ... + \frac{1}{10} + ...$$

Each time that Achilles traverses one of these spaces the tortoise traverses the following one. It may be objected, it is true, that the meeting point between Achilles and the tortoise can be calculated by the well-known formula giving the limit of the sum of an infinite number of terms of geometrical progression, of which the first term is a and the common ratio r is less than 1,

$$S = \frac{a}{1-r} \text{ that is } S = \frac{10}{1-\frac{1}{10}} = 11\tfrac{1}{9} \text{ yards.}$$

But, as Zeuthen [1] has pointed out, the very reasons appealed to by Zeno show that even in his time it was known how to effect this summation. What they disputed was precisely the legitimacy of the formula

$$S = \frac{a}{1-r},$$

[1] 29 Zeuthen, *Histoire des mathématiques*, p. 54.

since, in order to establish this, it is necessary, in

$$S = \frac{a}{1-r} - \frac{ar}{1-r}$$

to neglect the term $\frac{ar^n}{1-r}$ as insignificant. Up to what point is it right to do this? That is the question.

Instead of admitting the possibility of an infinite division, let us suppose that this division has a limit and that there exist ultimate elements of space, time and motion (whether in finite or infinite number, it matters little). To this Zeno replies with the paradox of the arrow. The extremities and the body of an arrow in flight must coincide at each instant with the points which compose its trajectory; but if there be a coincidence for however brief an instant of time, there is immobility. Then the movement of the arrow is reduced to a sum of instantaneous immobilities, which is absurd. If we attempt to avoid this objection by affirming that each instant corresponds, not to a certain position of the arrow, but to the passage from each position to the next, Zeno appeals to the argument of bodies which moving inversely to one another cross one another's paths, and he shows that the speeds supposed to be different are in reality equal, since by dichotomy the sum of the instants of which these speeds are composed can always be reduced to the same number, that is to infinity.

The arguments of Zeno in fact amount to the proof by *reductio ad absurdum* that a geometrical body is not a sum of points, that time is not a sum of instants, that motion is not the sum of passages from one point to another. They had the result of establishing once for all the infinite divisibility of space. Henceforward the discussion relating to divisibility dealt with matter,

THE MATHEMATICAL SCIENCES 131

and atomism could take shape, thanks to the work of Leucippus and Democritus.

From a mathematical point of view the problem to be solved is the following: no longer to identify discontinuous number with continuous magnitude, and yet to find a means of adapting number to the study of geometrical figures. This problem is difficult, for the reasoning of Zeno seems to be faultless, and the impossibility of reconciling it with the data of spatial intuition seems to condemn for ever the rational and direct use of mathematical infinity. On the other hand, in practical applications, certain sophists such as Antiphon affirm, on the basis of these reasonings, an identity between curvilinear and rectilinear elements which is inacceptable.

Thus, in spite of the efforts of Aristotle to render legitimate the notion of continuity, the confidence of Greek mathematicians in directly infinitesimal speculations was for ever shaken. Besides, the formulæ enunciated by Aristotle were not of any practical use in mathematics; they belonged to a treatise on physics which had in the highest degree a metaphysical character. To Aristotle, indeed, the question which presented itself is the following: "If infinity be a given reality, the enumeration of all the whole numbers must have a limit, which is logically impossible (*Phys.*, 204 b 4-10). But to reject infinity is to declare that time has a beginning, that magnitude is discontinuous and that the power to reckon has a limit (*Phys.*, 206 a 9-12). To remove these difficulties it is necessary, according to Aristotle, to distinguish between magnitude and number in the problem of infinity.[1] An infinite magnitude could no more exist than an infinite space. As a matter of fact, space could not extend beyond the material world of which it forms the boundary (*Phys.*, 212-31). If the universe were

[1] Cf. 21 Milhaud, *Etudes*, p. 120.

unlimited, it would not be able to accomplish its daily revolution in 24 hours. Further, what is infinite is imperfect, unfinished, and unthinkable ; yet the world is a finite whole which can be conceived in the mind.

But if magnitude be not infinitely great, it is *per contra* infinitely divisible, and, in this sense, there is an infinity of magnitudes, but only potentially and not actually, since the division is never completed. Continuity must be defined thus : that which is divisible into parts which are always divisible (*de coelo*, 268 to 6). If this be so, the arguments of the Eleatic school against the reality of motion lose all their force, for it is not necessary that the possible divisions of time and space should be performed in order that motion may really take place.

With regard to number, Aristotle adopts a quite opposite attitude. He admits the virtual existence of an infinite number, in this sense that after each whole number there is always another. But a numerical infinitely small is inconceivable, since unity is an element below which it is impossible to go.

To sum up, Aristotle considers all magnitude as finite, but he admits its infinite divisibility, thus rejecting spatial atomism. On the other hand, he affirms the extensible infinity of number, but not its infinite divisibility.

We see that though the views of Aristotle have undeniable metaphysical interest, they do not present any method of symbolizing and using, mathematically, continuity and infinity. From this point of view the problem discussed by Zeno remained untouched.

In order to avoid running counter to this problem, Greek science, with Eudoxus, had recourse to stratagem. This geometer begins by enunciating a theory of proportions which, taking into consideration geometrical continuity, is applicable to all ratios of magnitude, whether commensurable or not. If (A, B) and (C, D) be two pairs of magnitudes, the proportions A/B and C/D will be equal, if, whatever may be the whole numbers m and p, we always have

$$\frac{mA}{pB} = \frac{mC}{pD}$$

In this way the ratios of magnitudes become geometrical, and no longer simply arithmetical, as they had been to the Pythagoreans.

Having established this point, Eudoxus laid the foundations of an infinitesimal method by which it would be possible to pass gradually from a regular figure to the figure which circumscribes it. This method, called the method of exhaustion, is based on the following principles which are derived from the lemmas formulated for geometrical proportions.[1]

1. If two magnitudes a and b be unequal, the lesser repeated a sufficient number of times (n) will end by equalling or exceeding the greater. In other terms if $a < b$, $na \geqslant b$.

2. If from a magnitude there be taken more than its half, then from the remainder a part greater than half of this remainder, and so on indefinitely, there will be finally obtained a remainder less than any given magnitude.

It was by taking these principles as a basis that Eudoxus demonstrated, amongst other things, that circles have areas proportional to the squares of their diameter. The proposition is true for regular figures of 4, 8, 16, 32, etc., sides which are successively inscribed in the circles. Now, at each operation, the difference between the area of the circles and that of the new polygons inscribed is diminished by more than half. It tends to become zero, so that the properties established for polygons hold good for circles.

The method of exhaustion was taken up and given

[1] 26 Tannery, *Géo. grecque*, p. 96.

new life by Archimedes, who made successful and fruitful applications of it. Eudoxus had contented himself with showing by what lemmas a certain figure may be considered as the limit of another figure increasing progressively; but he did not know how to evaluate the successive terms of this progression. It was Archimedes who first discovered the practical means of effecting this calculation. He succeeded, for instance, in determining the circumference of a circle by defining it as the boundary of two polygonal perimeters, inscribed and circumscribed, of which the number of sides is indefinitely increased.

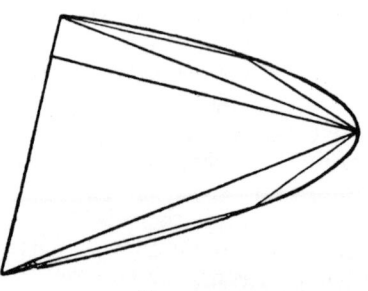

FIG. 14.

By an analogous process he was able to calculate curvilinear areas or areas bounded by curves. He showed that any segment bounded by a straight line and a parabola is equal to four-thirds of the triangle having the same base and the same height as the segment (Fig. 14). In this demonstration the passage to the limit is not directly used. In order to avoid this Archimedes proves that it would be absurd to suppose the area of the parabolic segment to be greater or less than four-thirds of the triangle having the same base and height.

The method of exhaustion rests on a *reductio ad absurdum* which proves its perfect logical exactitude.

THE MATHEMATICAL SCIENCES 135

This same exactitude prevented the Greek geometers from looking in another direction for the solution of the problem of areas and curvilinear volumes. By a stroke of genius Archimedes invented a method of integration based on the comparative study of the static moments of two figures, and which necessitates for this study the use of an infinite number of lines or parallel planes ; the comparison of suitably selected sections then gives the equation of equilibrium between the known surface or volume of one of the figures and the unknown surface or volume of the other. Thus to have equilibrium with a sphere, it is necessary to have four cones having as base the great circle and as height the radius of the sphere. The sphere has therefore a volume four times greater than that of the cone constructed with its radius. Archimedes, however, would not acknowledge any power of demonstration in this mechanical method, whose results, to be valid in his eyes, had to be confirmed by exhaustive reasoning. In fact, the Greek geometers considered that it was only by this reasoning that the dialectic of Zeno could be successfully refuted. On the one hand, the condition imposed on the difference (line or surface) of always diminishing by more than its half ensures that this difference can become less than any given quantity, after a finite number of operations. On the other hand, the method of construction employed in each problem ensures that the law of diminution is really obeyed by the decreasing magnitudes ; hence the terms which form the numerical representation of these constitute a series the convergence of which is evident and has no need of proof. In every way the direct use of infinity, which results from dichotomy, and which Zeno had criticized, is avoided.

However, the method of exhaustion thus understood remains difficult to manipulate. To make its application general, it would have been necessary to examine, as Cavalieri, Fermat and specially Pascal did later, the nature of the progressions which represent the decomposition of the geometrical figure. It would have been necessary to establish, once for all, the conditions which these progressions must satisfy in order to be used in the solution of any problem of quadrature. By following this path, the Greek geometers would perhaps have discovered some device similar to that used by Newton and Leibnitz, and they would have brought into their method a generalization of which they possessed the essential elements. But, being desirous above all to avoid the direct use of infinity, they were so intent on ensuring the rigour of the method of exhaustion in each particular case " that it left them no room to develop, beyond the need of the moment, the methods they employed to prove their results, or to create new methods."[1] Already necessitating lengthy demonstrations for relatively simple cases, the method of exhaustion became most complicated when used for the integration of surfaces and volumes of which the elements are connected by complex relations. So it is not astonishing that the successors of Archimedes, adhering to this method, were not able to carry on the brilliant work of their master, notwithstanding the time and knowledge at their disposal.

4. GEOMETRICAL ALGEBRA

Although the way opened up by Archimedes was but little followed, the comparative study of lines, surfaces and volumes nevertheless made real progress by means of what may be called geometrical algebra.

The Pythagoreans had already employed geometry in the study of the numerical properties of magnitudes regarded as commensurables, and thereby, as we have seen, they were restricted in spatial arithmetic.

[1] 29 Zeuthen, *Histoire des mathématiques*, p. 142.

The discovery of the irrational $\sqrt{2}$ dealt a first blow to this conception, which was completely shaken by the arguments of Zeno of Elea; but, before even the theory of proportions had been established by Eudoxus, the Greek geometers had succeeded in generalizing the quantitative study of magnitudes and in creating thus a kind of geometrical algebra. Their method was as follows:

The representation of a magnitude by the length of a segment can play the same part as the symbolical letters of algebra. This being so, in order to subtract or add two rational or irrational magnitudes, it is sufficient to represent them by segments, and then to place one of these segments on the other or on its extension.

The quantities which we call imaginary or negative certainly cannot be represented in this way; still, in many cases, the variations of the figure lend themselves partly to the same generalizations as the use of negative quantities in algebra.

As to the multiplication of magnitudes, in the direct sense, it is nonsensical, but it is possible to represent it indirectly by means of a rectangle whose sides are formed by the segments representing the two magnitudes to be multiplied.

In this manner a second geometrical expression of magnitudes is obtained, that is, as rectangular or square surfaces. To add or subtract them in this new form, it is necessary to give them a common side; one of the rectangles, whilst keeping the same area, is then transformed in such a way as to enable it to be applied exactly to the other. This operation is performed by means of the following proposition: the lines parallel to the sides of a rectangle, which intersect on one of the diagonals, divide this rectangle into four others, of which two are equal, that is, those which do not cross this diagonal (Fig. 15).

138   SCIENCE IN GRECO-ROMAN ANTIQUITY

For example, to add the rectangle B to the rectangle A of which one side is b, it is necessary to find a rectangle C (with sides b and x) which, being equal to B, can be applied to A by the common side b (Fig. 16).

To solve this problem it is necessary to proceed in the following way : On the extension of one of the sides of the rectangle B (Fig. 15) take the length b, then from the extremity of this side thus produced, draw the new diagonal to the point where it cuts the other side of B likewise produced. We have thus all the elements for constructing the rectangle C, which evidently fulfils the requirements of the problem and can be applied to the rectangle A. This construction is called παραβολή, or the application of surfaces. When made as we have just seen, it is *simple*, but it may be elliptic

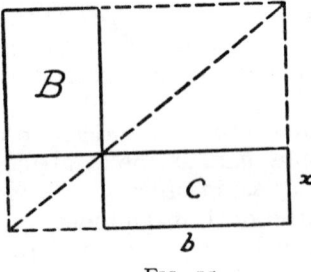

FIG. 15.

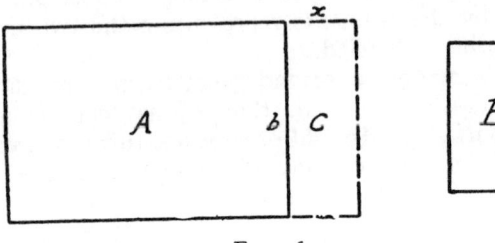

FIG. 16.

or hyperbolic. When *elliptic*, it corresponds to the following problem : on a given segment a construct a rectangle ax which when diminished by an unknown square x^2 is equal to a given square b^2 (Fig. 17).

In modern language the problem is expressed by the equation

$$ax - x^2 = b^2$$

or again, by adding and subtracting $\dfrac{a^2}{4}$,

$$\frac{a^2}{4} - \left(\frac{a^2}{4} + x^2 - ax\right) = b^2,$$

$$\left(\frac{a}{2}\right)^2 - \left(\frac{a}{2} - x\right)^2 = b^2.$$

The problem leads therefore to the construction of a

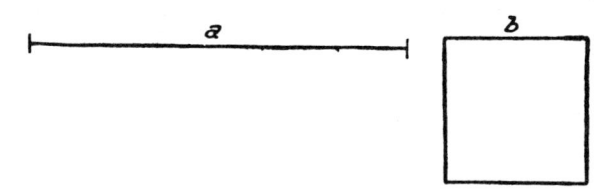

FIG. 17.

difference of squares. By putting the equation in the form

$$\left(\frac{a}{2}\right)^2 = b^2 + \left(\frac{a}{2} - x\right)^2,$$

the length $\left(\dfrac{a}{2} - x\right)$ and the length x are easily found by means of the theorem of Pythagoras.

Let a be the given segment and b the side of the given square. On one of the extremities of b, raise a perpendicular, then from the other describe an arc of circle of radius $\dfrac{a}{2}$ which will cut the perpendicular.

In this way we find the side $\dfrac{a}{2} - x$ and the length x (Fig. 18).

Once x is found, it is easy to construct the rectangle ax and the difference of the squares $\left(\dfrac{a}{2}\right)^2$ and $\left(\dfrac{a}{2}-x\right)^2$ (Fig. 19).

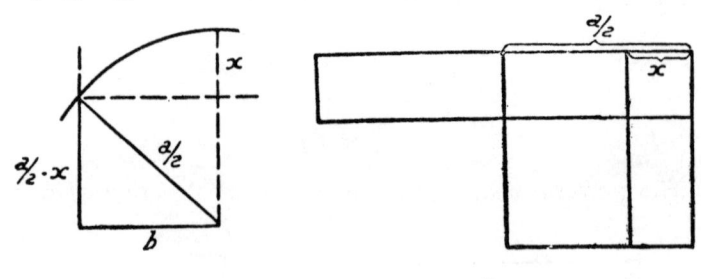

FIG. 18.   FIG. 19.

It can be seen that the rectangle ax, diminished by the square x^2 is equal to a gnomon whose surface is equal to the given square b^2 (Fig. 20).

The problem was afterwards generalized in the following manner: to determine two quantities of which the sum a is known and the product is con-

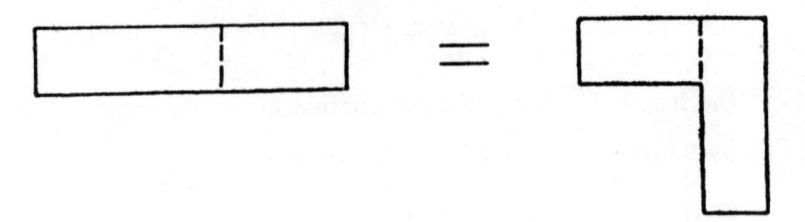

FIG. 20.

sidered as equal to a square b^2. To find the unknown value x one can proceed as follows: In a semi-circle of radius a inscribe the right-angled triangle of which b is the perpendicular dropped from the vertex of the right angle (Fig. 21). Under these conditions we have $b^2 = x(a - x)$ and when the roots of the equation are both positive they can immediately be found.

It can be seen that the treatment of magnitudes by geometrical representations is generally equivalent to their treatment by algebra. There is, however, a difference. Geometry is always fundamentally qualitative, while algebra is quantitative.[1]

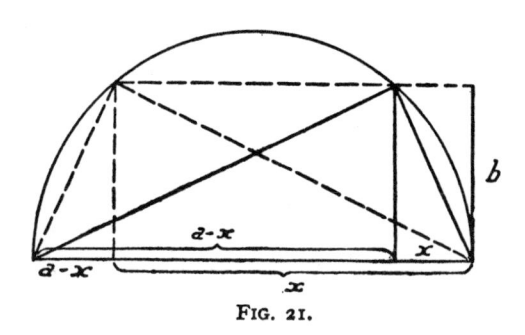

FIG. 21.

Whilst the elliptic application is by defect, the *hyperbolic* application is by excess and corresponds to the following problem: on a given segment a construct a rectangle ax which when increased by the unknown square x^2 is equal to a given square b^2. This problem is equal to the solution of the modern equation $ax + x^2 = b^2$ or by adding and subtracting $\dfrac{a^2}{4}$,

$$ax + x^2 + \frac{a^2}{4} - \frac{a^2}{4} = b^2,$$

$$\left(\frac{a}{2} + x\right)^2 - \left(\frac{a}{2}\right)^2 = b^2.$$

It is necessary, then, to construct as before a difference of squares. By means of the theorem of

[1] 4 Boutroux, *Idéal*, p. 74.

Pythagoras we quickly find $\left(\dfrac{a}{2} + x\right)$ and consequently x (Fig. 22). The rectangle ax is then easily obtained; the square x^2 is then added to it externally, instead of being taken away as in the elliptic application (Fig. 23).

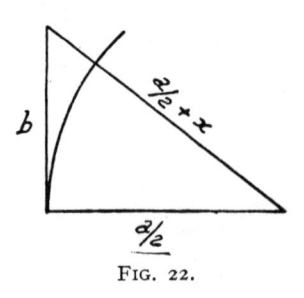

Without labouring the point, it can be seen that the Ancients have treated all the forms of the equation of the second degree which give positive roots; for them there could be no question of other roots, since they had no conception of them.[1]

The constructions which we have just mentioned are of no use when problems arise concerning the quadrature of the circle, the trisection of the angle and the duplication of the cube, which cannot be solved by means of the circle and the straight line. Recourse had then to be made to intercalations.

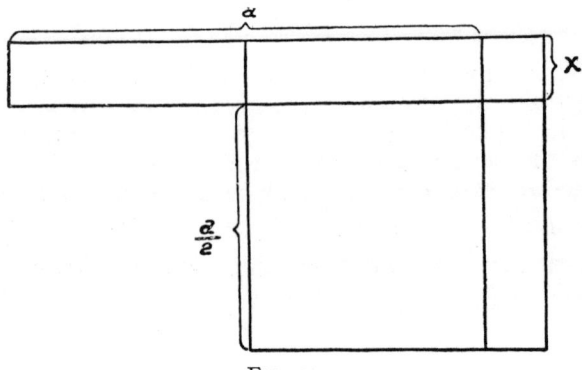

[1] 29 Zeuthen, *Histoire des mathématiques*, p. 39.

For example, to divide the angle ABC into three equal parts (Fig. 24). First AC is drawn perpendicular to BC and to AE which is parallel to BC ; then between AC and AE is intercalated DE = 2AB in such a way that its prolongation passes through B. F being the middle of DE and the triangle ADE being a right-angled triangle capable of being inscribed in a semi-circle of radius FE, we have radius AF = radius FE = AB by construction. The triangle ABF is

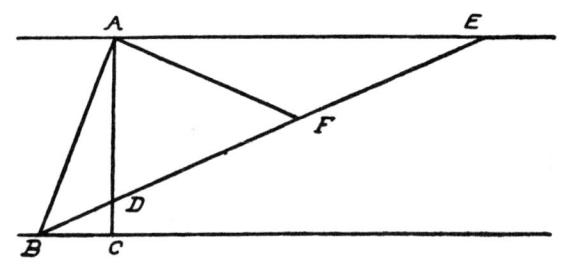

FIG. 24.

isoceles ; the angle ABF = AFB = twice the angle AEF = twice CBD. Hence

angle CBD = $\frac{1}{3}$ angle CBA.[1]

By intercalation must therefore be understood " the construction of a segment of a straight line of which the extremities are situated on given lines and which, when produced, passes through a given point. This segment can easily be obtained by means of a ruler (or piece of folded paper) in the following manner. On the ruler two marks are first made, the space between them being equal to the length of the given segment, then the ruler is turned round a fixed point and moved at the same time in such a way that one of the marks follows exactly one of the given lines. This

[1] Pappus, Hultsch Edition, Book IV, Prop. 32.

movement is continued until the other mark comes on the second given line."[1]

There was very probably a time when intercalation was admitted as a means of construction, together with the rule and compass, but it was soon rejected for reasons we have indicated (p. 119). It became then necessary to have recourse to conic sections, when the rule and compass were obviously insufficient. The consequence of the study of these sections was the development of the fruitful conception of "geometrical loci," for a conic section may be considered as the

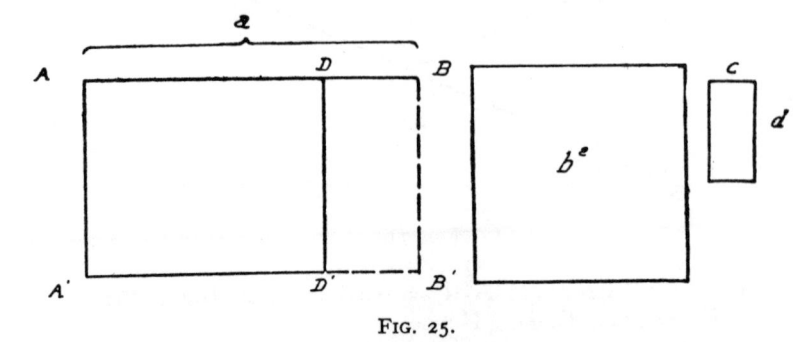

FIG. 25.

locus in which a cone and a plane meet. Hence there arose the expression of "*solid loci*," since the cone is a volume.

However, even in its most developed form, the theory of conic sections is closely connected with the first works on geometrical algebra. This is strikingly shown by the works of Apollonius.[2]

In these, the study of magnitudes and their ratios is always done by geometrical operations, only the field is enlarged thanks to the theory of proportions and similitude. This allows of the construction of surfaces

[1] 29 Zeuthen, *Histoire des mathématiques*, p. 66.
[2] 3 Boutroux, *Analyse*, p. 491 *et seq.*

which are similar (and not equal) to given surfaces.

For example, to construct on a given segment a a rectangle ax, which, diminished by a rectangle similar to a given rectangle cd, is equal to a given square b^2, we must have (Fig. 25):

$$\frac{DB}{c} = \frac{DD'}{d} = \frac{x}{d} \text{ whence } DB = \frac{c}{d}x.$$

AD is then equal to $a - \frac{c}{d}x$ and the unknown rectangle has for surface $x\left(a - \frac{c}{d}x\right)$; but as this must be equal to b^2, we have finally the equation of condition

$$ax - \frac{c}{d}x^2 = b^2.$$

The theory of proportions also enables the magnitudes which correspond to a given problem to be found

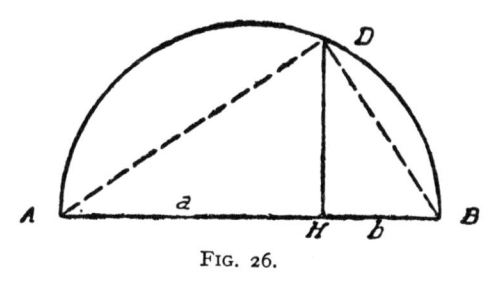

FIG. 26.

in a more direct manner. For example, to construct a square x^2, equal to a given rectangle ab, comes to finding a mean proportional between a and b, which is easy. Taking as diameter the segment AB of the length $a + b$ (Fig. 26), describe a semicircle, then at the extremity of a at H, raise a perpendicular $HD = x$. The triangle ADB inscribed in a semicircle is right-angled and we have $x^2 = ab$.

We can generalize the scope of this problem and say that the geometrical locus of the points D such that the perpendicular DH to AB is the mean proportional between the two segments which it determines on this straight line, is a circumference of diameter AB.

We can also, and it is here that conic sections intervene as geometrical loci, conceive of a more complicated relation of measurements; for instance, let us suppose that the segment AB being given, the segment DH is the side of a square subjected to the condition of being equal to a rectangle which, when applied to another

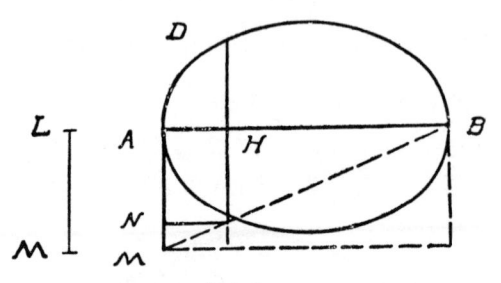

FIG. 27.

given segment LM, is at the same time *diminished* by a rectangle similar to the rectangle of dimensions LM and AB (Fig. 27).

To find any point of the locus, on the given segment AB, erect at its extremity a perpendicular AM equal to the second given segment LM. Construct the rectangle of the dimensions AB and AM, having MB as diagonal. From any point H draw a parallel to AM; this, at the point where it cuts the diagonal MB, determines the rectangle which, similar to the rectangle AB × AM, must be taken away from the rectangle of the dimensions AM and AH, applied to the segment AM (= LM). Then there only remains to find the side DH of the square equal to the rectangle AH × AN.

THE MATHEMATICAL SCIENCES

It can be demonstrated that the locus of the points satisfying the enunciation of the problem is an ellipse. If $AB = 2a$, $LM = 2p$, $AH = x$, $HD = y$, we shall have, according to the equation of condition (p. 145),

$$y^2 = 2px - \frac{2p}{2a} x^2 \text{ or } y^2 = 2px - \frac{p}{a} x^2.$$

When the rectangle to which the square DH^2 is equal is to be *increased*, instead of diminished, by a rectangle similar to the rectangle of dimensions AB

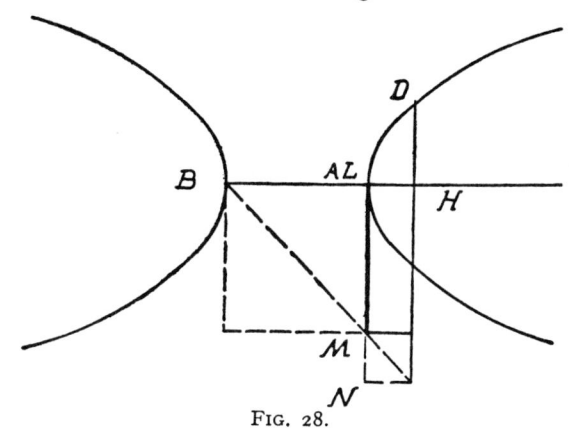

FIG. 28.

and LM, the geometrical locus is no longer an ellipse, but a hyperbola (Fig. 28).

Finally if the rectangle is not to be either diminished or increased but simply applied to the segment LM, we have the parabola.

Apollonius was of the opinion that whatever the conic section considered might be, the segment LM must always be perpendicular to the extremity of the segment AB even if the half chord HD be oblique in respect to the diameter AB. Hence the name of *latus rectum* (right side), which was given to it. For this reason, geometrical algebra renders the same services as those rendered later by analytical geometry. "Whilst we now express the fundamental property of a curve by an algebraical equation, Apollonius represented it by a figure; and owing to the fact that this auxiliary figure is drawn at right angles to the axis of abscissæ, even when the ordinates cut this axis at another angle, it always remains in some degree independent of the figure for the study of which it is used.[1]

Another fact, no less remarkable, was brought to

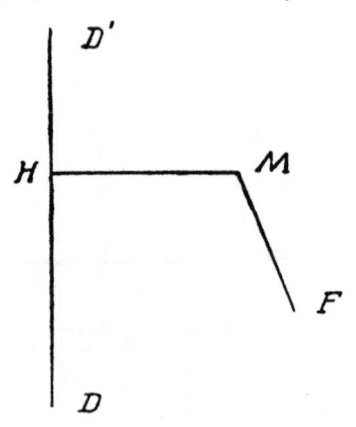

FIG. 29.

light by the Greek geometers (Pappus, Hultsch edit., book vii, prop. 238). Given an infinite straight line DD' (Fig. 29) and a point F, it can be demonstrated that the geometrical locus of the points M such that the ratios of the distances MF and MH from M to the point F and to the straight line be constant and equal to a given number, is a conic section. Inversely, being given any conic section, it is always possible to find a straight line and a point F which will allow the ratio in question to be established with regard to each point

[1] 29 Zeuthen, *Histoire des mathématiques*, p. 168.

of the curve. Further, according as this constant ratio $\frac{MF}{MH}$ is smaller than, greater than, or equal to 1, the conic is an ellipse, a hyperbola or a parabola.[1]

It is useless to enter into the details of the demonstrations, our aim being merely to show that the geomet-

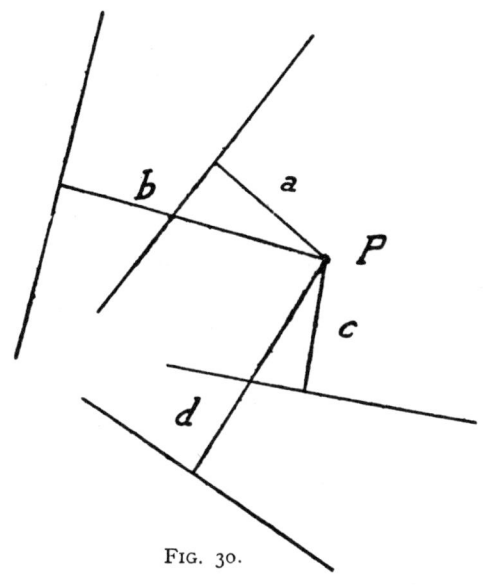

Fig. 30.

rical algebra of the Greeks, even in their most perfect works, remained faithful to its primary inspiration.

Let us add also that it was owing to conic sections that the study and investigation of loci was generalized. Among the problems considered by Pappus there are a number of this kind: from a point P drop the perpendiculars a, b, c, d on four straight lines. Find the locus of the points P such that the rectangle ab may be equal (or similar) to the rectangle cd (Fig. 30).

[1] 3 Boutroux, *Analyse*, I, p. 250.

The same problem may be stated in respect of six straight lines; the given ratio then relates no longer to areas but to volumes. The search for the geometrical locus then becomes very difficult by means of the methods known to the Ancients. Beyond six straight lines, they could not conceive that the problem could even be considered (Pappus, Hultsch edit., p. 680, 14). We know how Descartes by the help of analytical geometry surmounted the difficulties which had arrested their progress, and how he succeeded in solving in its generality the problem stated by Pappus.

5. THE ELEMENTS OF EUCLID—METHODS OF DEMONSTRATION—AXIOMS AND POSTULATES

It was not without difficulty that the Greek philosophers began to realize the rational structure of mathematics. As Proclus says, "It is difficult, in every science, to choose and to arrange in suitable order the elements from which and to which all the remainder proceeds. Of those who have attempted this, some have enlarged their collection, others have diminished it; some have used abridged demonstrations, others have lengthened indefinitely their demonstrations; some have avoided the reduction to the impossible, some, proportions; others have imagined preliminary developments in opposition to those who reject first principles; in a word, the various authors of *Elements* have invented a number of different systems.

"In such a treatise, it is necessary to avoid all that is superfluous—it is an impediment to the student; to bring together what is connected with the subject—an essential point for Science; to aim chiefly at clearness and conciseness—for their opposites perplex the intelligence; to seek to give the most general form to theorems—for the detail of instruction in particular cases only renders knowledge more difficult of attainment.

" From all these points of view, it will be found that the elementary treatise of Euclid surpasses any other : if its utility be considered, it leads to the theory of primordial figures ; [1] its lucidity and regular chain of reasoning are ensured by its progression from the most simple to the most complex, and by basing the theory on common ideas ; the generality of the demonstrations, by the choice of the starting-point in the problems to be dealt with, in the theorems which set forth the principles " (Proclus, *Comm. Eucl.* I, p. 73, 15 *et seq.*).[2]

The elementary treatise of Euclid is indeed a model of truly rational science. It begins by a collection of primary propositions which are enunciated in such a way as to make them universally acceptable and which, although as limited in number as possible, are sufficient to secure the construction of the whole mathematical edifice. This construction proceeds from the simple to the complex by way of demonstration and resolution of problems. It begins by establishing the properties of the most elementary figures, then by their means it demonstrates the properties of more and more complex figures. In this way the work of synthetic geometry is accomplished, and this work must be logically unassailable.

In dealing with the primary propositions, the *Elements*, as they have come down to us, distinguish between definitions and hypotheses (postulates and axioms).

The definitions (ὅροι) define the meaning and limits of the concepts used. The postulates (αἰτήματα) demand that certain constructions (for example, to draw a straight line between two points) shall be

[1] Polyhedra composed of material elements.
[2] Quoted from 26 Tannery, *Géo. grecque*, p. 142.

granted as possible without requiring proofs. The axioms or common notions (κοιναὶ ἔννοιαι) are truths which cannot be demonstrated but are self-evident (for example, the whole is greater than the part). It appears, however, that Euclid only admitted two kinds of primary propositions, definitions and postulates, and that he classified as one or the other propositions which were afterwards called axioms. This question is of secondary importance; it is of greater interest to examine whether the primary propositions of the *Elements* are in agreement with the conditions laid down by Euclid himself, and whether, on the other hand, they satisfy the exigencies of the modern use of axioms. With regard to the first point, it must be noted that the form of the definitions often leaves something to be desired. Such is the definition of the straight line, the empirical origin of which is purposely concealed, thus rendering it obscure.[1] Further, certain definitions, such as that of the diameter, contain useless elements. If the diameter be defined as passing through the centre, it is superfluous to add that it divides the circle into two equal parts.

As to the relation of the *Elements* to the modern theory of axioms, the following statements may be made :

Firstly, the primary propositions must be compatible, that is to say, not contradictory to each other, otherwise the consequences deduced from their combinations would necessarily be contradictory. The *Elements* fulfil this condition without proving it theoretically.

Secondly, the enunciation of a primary proposition must be rigorously complete. When we say that the whole is greater than the part, we must add, which Euclid has not done, that such an enunciation only concerns finite magnitudes and numbers. We know, in fact, that in infinity the part is equal to the whole;

[1] See page 119.

THE MATHEMATICAL SCIENCES 153

for instance, the summation of the series of even whole numbers is equivalent to the summation of all whole numbers, since between the terms of these two summations we can establish a univocal and reciprocal correspondence. It is easy to verify this by writing the two series as follows:

$$1 \quad 2 \quad 3 \quad 4 \ldots n \ldots$$
$$2 \quad 4 \quad 6 \quad 8 \ldots N \ldots$$

To every whole number a corresponding even number can be found *ad infinitum*.

Thirdly, the primary propositions must be in sufficient number, without any being superfluous. The *Elements*, in spite of their endeavour to be complete, sometimes leave much to be desired in this respect. Often they omit to justify by an axiom facts regarded as evident, even when they are not derived from the principles primarily postulated; for example, the following statement: if A, B, C be three points belonging to the same straight line and if B be between A and C, it will also be between C and A.[1]

Finally, it is essential that the primary propositions considered necessary for the building up of geometry should form a logically indissoluble whole, that is composed in such a way that not one part can be suppressed or altered without involving the ruin of the whole edifice. If the suppression or change of one of the primary propositions should lead to consequences which, without being logically absurd, were simply different from what they were before, the necessary conclusion would be that various types of geometry are equally possible, that is to say equally true from a logical point of view.

This problem did not present itself to Euclid; but he has intuitively understood its importance, by claiming as a postulate that from a point taken outside

[1] 5 Boutroux, *Les mathématiques*, p. 73.

a straight line only one parallel can be drawn to it. Seeing the hypothetical character which he gives to this proposition, Euclid has had regard to the exigencies of the modern theory of axioms, but if, as he believed, only one geometry is possible, his hypothesis would appear strange and superfluous, for one would necessarily be able to affirm the singleness of the parallel and deduce from it the definitions already postulated of the straight line, the plane and angles. It would seem that he must speak of a theorem of parallels and not of a postulate if logically there only exists but one geometry.

The successors of Euclid were of this opinion, and not without reason, and this is why they endeavoured to demonstrate the proposition which Euclid had enunciated as a hypothesis, but all their attempts in this direction were in vain.

In the nineteenth century they surrendered to evidence. It is possible to abandon the postulate of the parallels, whilst keeping the other primary propositions. Geometries can then be constructed which have other properties than that of Euclid and which for this reason are called non-Euclidean (Lobatschewsky, Riemann). These geometries, the truth of which is guaranteed by logic, deal with mathematical facts (lines, surfaces, angles) which are real and in no wise fanciful, although we cannot picture them by intuitive perception. The field of geometry is therefore vaster than Euclid supposed, but although he did not entirely construct the modern theory of axioms, to him belongs the merit of having established it upon a permanent basis.

The primary notions having once been elucidated, it is possible by logical deduction to link to them a series of propositions entirely derived from one another. These propositions are classified and distinguished according to their nature. There is first the *theorem*

THE MATHEMATICAL SCIENCES 155

or principal proposition; then the *lemma*, a secondary proposition intended to facilitate the demonstration of a theorem to follow; and the *corollary*, a direct consequence of a theorem which has just been established.

But how are these propositions to be demonstrated? Although all agree as to the method to be followed, there is a divergence of views as to the interpretation to be given to the demonstration. At the time of Plato and probably of Euclid also [1] there were subtle discussions on the question whether mathematical propositions must be considered as problems to be solved, or on the contrary as theorems to be demonstrated. Proclus (*Comm. Eucl.*, I, p. 77, 15 *et seq.*) sums up the discussions on this subject in the following way. The Platonists such as Speusippus and Geminus held that figures and their properties exist in the eternal world of ideas independently of the construction the mathematician can make of them; the latter can only make manifest to the understanding what already existed. For example, equilateral triangles are such by definition, that is to say, by an eternal relation of ideas, and the fact of constructing them cannot add to or take away anything from their existence. Therefore it is not correct to speak of problems, but only of theorems (objects of contemplation). Some philosophers, such as the mathematicians of the school of Menaechmus, were of the opinion that all should be regarded as problems; others said with Carpus that problems as a class precede theorems, because it is by the former that the subjects are found to which belong the properties to be studied.

Finally, many considered as a theorem that which contained only one possibility, and as a problem that which was capable of several possibilities. For example, " to propose to inscribe a right angle in a semicircle is not to speak geometrically, since all the

[1] 26 Tannery, *Géo. grecque*, p. 145.

inscribed angles are right angles; on the contrary, to inscribe an equilateral triangle in a circle is really a problem, since it is possible to inscribe in it a triangle which is not equilateral."[1]

The disagreement is deeper in appearance than in reality, and arises, as Proclus explains, from a difference of point of view. The distinction between ideal science and didactic science is itself sufficient to show that both Geminus and Carpus may be right, " for if it is according to the order that Carpus gives the pre-eminence to problems, it is according to the degree of perfection that Geminus gives it to theorems."[2] In as far as it is ideally conceived of, mathematical truth only contains theorems, but to the mind that conquers it by degrees it appears in the form of problems. However, whether it is a question of problems to solve or theorems to demonstrate, it is necessary to have recourse to methods of which the Greeks, starting from Plato, had carefully fixed the stages. By analysis they decomposed a complex whole into simpler propositions, already admitted or demonstrated. For example, to draw a tangent to two circles, they supposed the problem solved, and showed that in order to find this solution, it is necessary to start from the known construction of a tangent drawn to a circle through an external point. Synthesis, on the contrary, enables the complex geometrical relation, of which the demonstration is needed, to be reconstructed by means of primitive propositions.

For the Greeks the typical question consists of seven parts :

1. The *protasis*, or enunciation indicating the data of the problem and what is required ;
2. The *ecthesis*, or repetition of the enunciation in relation to a particular figure ;

[1] 26 Tannery, *Géo. grecque*, p. 145.
[2] Quoted according to 4 Boutroux, *Idéal*, p. 63.

THE MATHEMATICAL SCIENCES 157

3. The *apagogee* (ἀπαγωγή),[1] which changes the question propounded into another more simple;

4. The *solution*, which shows the possibility of solving this simpler question by means of the data of the enunciation in defining by *division* the conditions of possibility;

5. The *construction*, which completes the ecthesis by defining the various accessory lines which it is necessary to consider in order to make the demonstration;

6. The *demonstration* properly so called, which deduces from the construction the figure required;

7. The *conclusion*, which affirms that this figure satisfies the required conditions.[2]

As M. Zeuthen remarks, "whilst the *analysis* contained in Nos. 3 and 4, i.e. in the transformation and the solution, is methodically important for finding the solution, it is no longer necessary when it is merely a question of expounding in an unassailable manner what has been found, which was always the chief aim of Greek writers. It is therefore very often omitted, so that the exposition consists only of the use of operations numbered 1, 2, 5, 6, 7; thus the form which we call synthetic is obtained."[3] By their very nature theorems assume the form of a synthetical rather than an analytical exposition. They are capable, however, of an antithetical demonstration, the procedure of which is analytical. One supposes that the proposed theorem be true or false, then one considers whether the consequences deduced from this supposition be apparently right; according to the conclusion reached, the theorem will be judged true or false. One supposes, for example, that two triangles, having one side and

[1] G. Friedlein, *In primum Euclidis Elementorum librum Procli Commentarii*, Teubner, Leipzig, 1873, p. 212.
[2] 4 Boutroux, *Idéal*, p. 55.—29 Zeuthen, *Histoire des mathématiques*, p. 80.—26 Tannery, *Géo. grecque*, p. 148.
[3] 29 Zeuthen, *Histoire des mathématiques*, p. 83.

two angles adjacent to this side equal each to each, are equal. To affirm the contrary would be to admit that the two triangles cannot be exactly superposed, and that the angles supposed to be equal are not so in reality, which is not in agreement with the data of the question.

If we now consider Greek geometry, having no longer regard to its particular methods, but to its spirit, there are other characteristics yet to be noted. The demonstrations are always instinctively based on logical and statical ideas ; they generally avoid making any appeal to considerations which, in spite of their evidence, arise from intuitive perception. It is thus that Euclid demonstrates the following fact which might appear however unquestionably evident : if from a given point a perpendicular and two oblique lines are let fall on a straight line, of those two oblique lines that which diverges most from the perpendicular will be the longer.

As far as possible Euclid also avoids, if not the displacement, the turning over of a figure, although this operation, now considered correct, allows of a more rapid demonstration. For instance, it is enough to turn over an isosceles triangle in order to demonstrate that the angles opposite to the equal sides are themselves equal. Euclid however prefers to decompose the isosceles triangle into two right-angled triangles, whose equality he then proves. It is the same when he wishes to demonstrate, pair by pair, the equality of the angles formed by a secant which cuts two parallel straight lines. The simplest method would be to displace one of the parallels until it coincides with the other. Euclid here again brings in two right-angled triangles, of which he establishes the equality. In this way the demonstration preserves a static character more in agreement with the exigencies of logic. This is so true that wherever displacement occurs in plane geometry, it is equivalent to a construction. Thus to superpose a triangle B on another triangle A in such a way as to be able to compare them, comes to constructing the triangle B on the triangle A according to the conditions stated in the enunciation.

We see that plane geometry avoids the direct use of the methods of displacement, especially of turning over, and the reason for this must be sought in the fear of giving a hold to the arguments of Zeno respecting motion and infinity.

It was also for this same reason, we think, that the Greek philosophers avoided the geometrical infinity in the same way as they rejected the direct use of numerical infinity in their methods of integration. They possessed, however, since the works of Apollonius, the essential elements (points of involution, anharmonic ratio) for reaching, by generalization, to geometrical infinity. But on this question they remained faithful to the teaching of Aristotle, who considered real space, and therefore geometrical space, to be finite. Consequently, the conception of points, straight lines, and planes, removed to infinity, is not only obscure from a logical point of view, but contrary to experience. Therefore it would not be possible, even as a convenient symbolism, to appeal to geometrical infinity and make it the starting-point of new methods. For want of searching in this direction and from loyalty to its logical ideal, Greek geometry was obliged to resort to a complicated kind of demonstration, the application of which rendered difficult the linking of theorems in correct sequence. It was an event of outstanding importance when Desargues, in the seventeenth century, made a direct use of geometrical infinity. The simplifications wrought by this act were so great that they struck the contemporaries of the great geometer. Speaking of Desargues, the engraver Bosse says that the work which he has published on conic sections, one proposition of which includes as consequences sixty of those of the four first books of Apollonius, has gained for him the esteem of savants.[1]

In conclusion, what characterized the spirit and methods of Greek geometry was an ideal of logical rationality which may be defined in the following terms:

1. To postulate primary propositions (definitions, hypotheses) as logical and as few in number as possible.

2. To construct by means of reasoned deduction the whole edifice of mathematics on the basis of these propositions.

Logical rigour is thus safeguarded, but at the price of complications which, as we have just seen, do not allow the methods of invention and demonstration to be given all the generality of which they are capable.

[1] Chasles, *Aperçu historique des méthodes*, Gauthier-Villars, Paris, 1875, p. 78.

CHAPTER II

ASTRONOMY

FROM its beginnings Greek Astronomy, like Geometry, sought to model itself after the type of a rational science; having to explain physical facts, it tried to do so by physical causes, that is to say causes of the same nature as these facts.

To primitive peoples, celestial phenomena are divine, that is, they depend entirely on the more or less capricious will of divinities. Doubtless, as we have seen, the Egyptians and Chaldeans already possessed some amount of astronomical knowledge, but this knowledge consisted, after all, in ascertaining the periodicity of celestial phenomena, without giving any explanation of these.

From the first, Greek astronomy launched out in another direction, as the works of the Ionian school show. These works appear incredibly daring if we compare them with the religious beliefs of the Chaldeans and Egyptians.

Thales, for example, lays down as a principle that water is the unique element from which all things arise by the action of purely physical causes, for water can be solidified into ice, be changed into vapour, that is, air, etc. Having once laid down this principle, Thales deduces from it a cosmology which, in spite of its childish simplicity, remains physically rational.

However it was only with difficulty that Greek astronomy succeeded in specifying its ideal and object. It passed through a series of stages which may be roughly indicated as follows: in the first phase astronomy is entirely confused with meteorology; in the second, the physical and geometrical hypotheses which it needs are distinguished more or less clearly; in the third and last phase an attempt is made to give a mathematical representation as exact as possible of the movement of the heavenly bodies.

1. METEOROLOGICAL IDEAS

As long as the earth and the sky were regarded as being situated on the confines of one another, celestial phenomena were assimilated to meteorological phenomena and an explanation of the former was sought in the latter. The meteorological ideas themselves were very confused. Vapour was simply condensed air. Furthermore up to the eighth century B.C. darkness was considered as a material thing, composed of vapour. Heraclitus, for instance, affirms that darkness is a concrete vapour, which, rising from the sea and the bottom of the valleys, is able by its aqueous nature to extinguish the sun. Plato likewise makes the Pythagorean Timaeus say that fog and darkness are condensed air (*Timaeus*, 58 D, 2). The air possesses different properties according as it is hot or cold : in the first case it is light and mobile; in the second it is heavy and stable. On the other hand, when it is compressed in the form of vapour, it is partially changed into invisible fire which suddenly bursts forth as lightning, when, for lack of compression, the cloud is rent. For a long time the Greeks, like the Chaldeans and the Hebrews, regarded daylight as distinct from sunlight. Shadow even had a concrete reality of its own, it was not a function of light; it was only strengthened by its opposition to light. These ideas persisted until the time of Empedocles, when the reflection of light and the true nature of vapour, shadow, and darkness were discovered.

ASTRONOMY 163

In agreement with the meteorological opinions which we have just called to mind, there were, concerning the nature and the movement of the heavenly bodies, the eclipses, the shape and position of the earth, very diverse hypotheses of which the following are the principal.

First of all, to explain the constitution and the movements of the heavenly bodies, Thales and with him Heraclitus considered them as basins which move on the liquid vault of the heavens and in which the dry exhalations arising from the earth are consumed. Anaximander and probably with him Pythagoras likened the celestial bodies to the felloes of a wheel, which, formed by the compression of the air, encloses an invisible fire; owing to the compression, openings by which the fire escapes are produced on the periphery of the felloes, which revolve with a uniform movement.[1] Anaximenes, on the contrary, declares that the celestial bodies are of an igneous nature and are supported by the air " like thin leaves." [2] Xenophanes considered them to be fiery clouds, similar to St. Elmo's fire, which move in a straight line from east to west.[3] Empedocles thought, as we have seen, that the sun was produced by the rays which proceed from the lighted hemisphere and which, after being reflected on the surface of the earth, are concentrated at one point of the crystalline vault. Anaxagoras appears to have been the first to describe the sun, the moon, etc., as fiery stones which are drawn round by the rotation of the ether.

The explanation of eclipses arises quite naturally from these various ideas. In the cosmology of Thales and Heraclitus, the eclipses, according as they are partial or total, are caused by the inclination or turning over of the luminous face of the basins which

[1] 8 Burnet, *Aurore*, pp. 68 and 124.
[2] *Ibid.*, p. 31.
[3] *Ibid.*, p. 135.

are the stars. According to Anaximander, they result from the partial or total obstruction of the opening of the felloes. Anaximenes explains them by the intervention of earthy dark bodies which move around the celestial vault. Empedocles, however, knew the true theory of the solar eclipses, though it was Anaxagoras who clearly formulated it, as Hippolytus reports: "The moon is eclipsed by the earth which robs it of the light of the sun, and also sometimes by the bodies which are below it and pass in front of it. The sun is eclipsed at new moon, when the moon hides it from us." (Diels, *Vor.* I, 301, 47.)

As to the shape and position of the earth, the first Ionians generally considered this as a cylinder supported by water or suspended in the air, or as a thin disc, or again as a dish with turned-up edges. Pythagoras seems to have been the first to affirm the sphericity of the earth, which was distinctly proclaimed by Parmenides.[1]

Finally, it may be said that the conceptions of the comparative movements of the heavenly bodies are lacking in precision, and vary according to their authors. These all agree that the region of the fixed stars accomplishes a revolution round the celestial pole in 24 hours; but they differ in their views regarding the sun, moon and planets. These heavenly bodies are sometimes regarded as meteors which traverse the atmosphere by an independent motion, sometimes as bodies partially drawn by the movement of revolution of the starry heaven.

2. THE PHYSICAL HYPOTHESES

The Pythagorean school did not entirely abandon the meteorological studies of its predecessors, but it added to them the desire to comprehend the mechanism

[1] 25 Tannery, *Science hellène*, p. 208.

of the celestial movements. In the doctrines professed by this school, it is very difficult to separate the ideas of the master from those of his disciples.

Although Pythagoras affirmed that the earth is motionless,[1] it appears that he must be given the credit of recognizing that it is a sphere, it may be because he considered this figure perfect, or it may be that he had recognized it in the shape of the terrestrial shadow which causes the lunar eclipses. He was the first to distinguish, in the progression of the sun, of the moon and even of the planets, two movements which take place about distinct poles. One of these movements is diurnal, along the plane of the equator; the other is annual, in an opposite direction to the first, along the plane of the ecliptic. This is all that can be reasonably attributed to Pythagoras.

One of his disciples, Philolaus, a contemporary of Socrates, developed the conceptions of his master in the following manner. The spherical universe is surrounded by a fire which sustains it, and of which a part is also condensed at its centre. The central fire produces the diffused light of day and the outer fire feeds the stars. The space which separates them is divided into three concentric regions. The most distant is the *Olympus*, or the sphere of the fixed stars. Then comes the *Cosmos*, in which are found successively, as the central fire is approached, the planets, the sun, and the moon. The sun, moreover, is not self-luminous, it is a transparent mass like glass, which receives the illumination of the fire from above and sends it back to the earth. Lastly, the *Uranus* forms the sublunar region in which " are found the things subject to generation, the prerogative of that which animates the transmutations." (Aetius, Diels, *Vor*. I, 237, 23.) This radical distinction between the sublunar region and the space which extends from the moon to the confines

[1] 13 Duhem, *Système*, I, p. 8.

of the universe was revived by Aristotle and affirmed until the Renaissance.

The bodies which exist above the moon are composed of pure fire or pure elements, which cannot be impaired or changed; they are therefore eternal, and, being uncreated, are imperishable.

The sub-lunar bodies, on the contrary, are all complex; they are subject to generation and destruction, since the mixtures of which they are formed are subject to all sorts of changes.

The earth is in the Uranus as well as its opposite the counter-earth (Antichthon), which was postulated to satisfy the law of perfection which required that the number of the heavenly bodies in circular motion should reach the perfect figure ten. The existence of the counter-earth was also necessary to explain the greater frequency of eclipses of the moon than of the sun.

The earth and the counter-earth turn around the central fire as if they were rigidly fixed to the extremities of one diameter. This is why we cannot see either the central fire or the counter-earth from the side on which we live. The ten celestial bodies (sphere of the stars, five planets, sun, moon, earth and counter-earth) move around the central fire, the hearth of the universe, after the manner of a chorus on the stage; moving at different speeds, they produce by their revolution a perfect musical harmony. The earth is not the only heavenly body inhabited. The moon is also inhabited, but the lunar beings are more beautiful and fifteen times as big as the terrestrial beings. (Aetius, Diels, *Vor.*, 237, 43.)

Hicetas and Ecphantus, two disciples of Pythagoras later than Philolaus, abandoned the hypothesis of the counter-earth; they placed the central fire in the interior of the earth and the earth itself at the centre of the universe. Furthermore, to explain the move-

ASTRONOMY

ment of the heaven and the heavenly bodies, which movement they considered as being apparent, they endowed the earth with a movement of rotation on itself. Their doctrine, preserved by Cicero amongst others, certainly guided Copernicus in his investigations, for he twice quotes the passage from Cicero (*Quaestiones Academicae priores*, II, 39), in which Hicetas is erroneously called Nicetas. This passage is as follows:

"According to Theophrastus, Nicetas of Syracuse professed the opinion that the sun, moon and all the celestial bodies remained motionless, and that nothing moves in the world, except the earth, which, turning round its axis at a great speed, produces the same appearances as those observed when it was supposed that the earth was fixed and the heaven in motion. Some think that Plato, in the *Timaeus*, said the same thing in a somewhat more obscure manner."[1]

As M. Duhem remarks,[2] the little that we know of the systems elaborated by the Pythagoreans to explain the celestial movements is enough to awaken our astonishment and admiration. The fecundity and the ingenuity of the Hellenic mind are surprising: scarcely had it found itself at grips with the astronomical problem when it multiplied its attempts at solution, and attacked it in most diverse ways. The conceptions of the Pythagorean school had in fact an incalculable influence on astronomy, for they distinguished for the first time between movements which are real and movements which are only apparent; they bring into relief the fact that outside the data immediately furnished by the senses there must be sought a harmonious reason to explain them.

Plato incorporates in his teaching the principal elements of the Pythagorean astronomy. He retains

[1] Quoted after 13 Duhem, *Système* I, p. 22.
[2] *Ibid.*, p. 27.

the fundamental distinction between the diurnal motion and the annual retrogradation of the planets, sun and moon, movement and retrogradation which take place on two planes and about two different poles. The *Timaeus* shows us the Demiurge who, after having created a world-soul, cuts it in the shape of a X, then curves back the extremities of this X so as to obtain two circles. One of these circles represents the equator and the uniform changeless movement of the diurnal revolution ; the other represents the ecliptic and the varied movements of the celestial bodies other than the stars.

The two circles are found again in the movements of the mind, which sometimes seeks after the eternal, sometimes, on the contrary, clings to the changing elements of reality. But the principal idea of the Pythagorean astronomy, which Plato kept, was the opposition between real and apparent movements. For this reason, he assigns to astronomy the following task : to account for these appearances, that is, to discover behind the sensible phenomena the geometrical reasons which explain and justify them. " Plato, says Simplicius in his Commentaries (in *Aristotelis libros de coelo commentarii*, Bk. II, cap. xii, Karsten edit., p. 219, col. a), admits in principle that the celestial bodies move with a circular motion, uniform and constantly regular (that is, in the same direction) ; he propounds therefore this problem to mathematicians—What are the circular and perfectly regular movements which may properly be taken as hypotheses to account for the appearances of the wandering heavenly bodies ? " [1]

The problem having been thus stated, it is necessary, starting from Plato, to distinguish in Greek astronomy two kinds of hypotheses which until that time had been more or less mingled : the physical hypotheses regarding the nature and constitution of the stars, and

[1] Quoted from 13 Duhem, *Système* I, p. 103.

ASTRONOMY

the mathematical hypotheses which attempt to account for their movements. The physical hypotheses, although in some degree supplemented by Aristotle, remained in antiquity and the Middle Ages practically the same as in the time of Plato and his immediate predecessors. At this epoch, as we have seen, the air and humidity were no longer confused; darkness was considered as a shadow and no more as a material fact. It was also admitted that although the sun, planets and stars shine by themselves, the moon has a borrowed light.

This being so, the physical hypotheses may be reduced to four:

1. The universe forms a *finite* and finished whole. To suppose it illimitable, is to contradict both reason and fact. Our reason cannot in fact conceive of something which exists in reality and does not occupy a definite place. On the other hand, if the universe were infinite, its extremities would have to traverse infinite spaces in a finite period of twenty-four hours, which is actually impossible.

2. Since the universe is finite, it has a spherical form and a *centre*, and it is the earth which must occupy this centre. If we consider the earth alone, we see that it is motionless. Besides, of all the elements known to us, it is the terrestrial element which is heaviest and consequently must occupy the centre of the universe.

3. The universe as a whole is composed of two regions: one *celestial*, the other *sublunar*. The sublunar region comprises the bodies formed by the mixture of the four elements, water, air, earth and fire, and which are therefore subject to birth and death. The celestial region is occupied by the heavenly bodies, which, being formed of a fifth and unique element (quintessence), are, like this element, incorruptible.

4. Physically there is but one possible movement,

170 SCIENCE IN GRECO-ROMAN ANTIQUITY

regular and uniform, for a body which turns freely about another; it is the *circular* movement. For, if the revolving body begins to approach or recede from the central body, it will end either by falling on it or by going away from it altogether.

3. THE MATHEMATICAL HYPOTHESES

The mathematical hypotheses were on the whole much more varied than the physical ones. In order to grasp their significance, it must be remembered that they do not pretend to explain the movements of the heavenly bodies in regard to one another by any physical cause such as Newton's law of attraction, for instance.[1] They only attempt to give a geometrical representation of these movements. This representation may be imaginary like the mechanical means for going from the earth to the moon imagined by modern novelists. The novelist must doubtless take into account the known laws of physics and not contradict these: but nevertheless it matters little to him that the engineer has not the necessary funds for the construction of the cannon which will send a bullet to the moon. In the same way Greek astronomy was obliged to take into consideration the four physical facts mentioned above, but for the rest it was entirely free to invent whatever geometrical representation appeared to be most appropriate. Plato and his Pythagorean

[1] "We must, however, except a curious opinion reported by Plutarch (*De facie in orbe lunae*, Ch. VI) which seems to foreshadow the mechanics of Newton, and which may be summarized as follows: What keeps the moon from falling is its own movement and the rapidity of its rotation; similarly, for a projectile put in a sling, the force which prevents it from falling comes from circular rotation. In fact, natural motion only carries along a given body if nothing else opposes it. The moon is not carried along by its weight, for this weight is repelled and destroyed by the force of its rotation." Quoted from Doublet, *Histoire de l'Astronomie*, p. 119.

ASTRONOMY

predecessors thought to account for the appearance of the wandering heavenly bodies by endowing them with a dual revolution, diurnal and non-diurnal, in an opposite direction to each other; but this conception did not solve the problem.

The planets situated on the same plane (the ecliptic) as the sun doubtless traverse the same region as the sun, namely, the constellations of the zodiac, but their progression is irregular and shows stationary points followed by a retrograde movement, then an advance, and so on.

To account for this irregular motion, Eudoxus of Cnidus gave to each wandering heavenly body a mechanism of homocentric spheres touching and enclosing one another and having the earth as their centre. " The heavenly body is situated in the thickness of the last of these spheres, the one which is within all the others, and its centre is on the equator of this sphere." [1]

The first sphere, that which is exterior to all the rest, turns with a uniform motion from east to west, in twenty-four hours round the axis of the earth shown by the Pole star. In this manner all the planets share in the diurnal rotation which moves the heavenly bodies. The second sphere, resting by means of its axis on the first sphere, is animated by the same uniform movement, but the speed and sense, as well as the direction, of its own movement are different. In fact, this second sphere turns uniformly from west to east around an axis which is normal to the ecliptic. The duration of this revolution is not the same for the various planets; it is, for example according to Eudoxus, one year for Mercury, eleven years for Jupiter, etc.

The third sphere, which is interior and contiguous to the second, is affected by the complex movement of

[1] The system of Eudoxus has been reconstituted by Schiaparelli and summarized by 13 Duhem, *Système* I, p. 114.

the latter and combines this with its own uniform movement.

Things proceed in this manner down to the last sphere, which carries the planet on its equator, and as many spheres are required as there are particular movements of the planet to explain.

For instance, if the plane of the moon were the same as that of the ecliptic, there would be as many eclipses of the sun and moon as there are new and full moons, and two spheres would be sufficient to account for the observed facts. But the plane of the moon being inclined to that of the ecliptic, the latter is cut by the lunar orbit at two points or nodes, at which points alone eclipses can take place. As these nodes are displaced by a uniform and regular movement, it requires a special sphere to explain this displacement. So that three spheres in all are necessary to explain the movement of the moon in the heaven.

The problem is more complicated where the planets are concerned, since here there are stationary points and retrogradations followed by new progressions. Thus for each planet Eudoxus had recourse to four spheres: the first is connected with the diurnal revolution, the second with the zodiacal revolution, the third and fourth with the irregular movements.

There would be in all 27 spheres (20 for the planets, three for the sun, three for the moon, and one for the stars).

Aristotle adopted the system of Eudoxus and sought to perfect it, partly by his own ideas and partly by those of Calippus. In the system of Eudoxus the movement of each planet forms an independent whole. Aristotle imagined compensating spheres which are intercalated in the spaces between the various mechanisms of the heavenly bodies. All the movements of the planets then become one with the single movement which animates the starry sphere. Aristotle

ASTRONOMY

also affirmed the materiality of the spheres by considering them to be composed of ether. This materialization was generally abandoned later on, until the time when the idea was revived by the Arabs.

The system of homocentric spheres, perfected by Calippus and Aristotle a hundred years before, survived until towards the end of the third century B.C. It clashed, however, with weighty arguments based on the noticeable variations of brightness shown by the planets, especially Mars and Venus. These variations of brightness indicate that the distances of the planets from the observer change in a manner which is incompatible with a system of spheres concentric to the earth, in which the planets are always equally distant from the earth.[1] Further, the theory of Eudoxus does not explain why Mercury and Venus are the only planets which always remain in the neighbourhood of the sun.

To surmount these difficulties, a disciple of Plato, namely Heraclides of Pontus, had recourse to two hypotheses, of which one, which is quite original, admits a partial heliocentrism. Like the Pythagorean Ecphantus he first of all affirmed that the earth is at the centre of an infinite universe and that it turns on its axis in twenty-four hours, which explains the apparent revolution of the starry heavens. This being so, he supposed that Venus and Mercury revolve round the sun, whilst the latter moves round the earth as do the other planets.

Aristarchus of Samos, the date of whose scientific work is about 280 B.C., went farther still in the same direction. He conceived a heliocentric system, the essential ideas of which were reproduced by Copernicus in the sixteenth century, and which may be described as follows: the motionless sun is situated at the centre of the universe which is bounded by the immobile

[1] 2 Bigourdan, *Astronomie*, p. 254.

sphere of the fixed stars. The earth is animated by a dual movement: the diurnal movement of rotation on its axis, and the annual movement of revolution round the sun. The planets also revolve round the sun. According to Aristarchus it must also be supposed that the sphere of the stars is very far away, otherwise the existence of parallaxes would be ascertained, which, in his opinion, is not the case.

This conception, as ingenious as audacious, had no renown in antiquity. The reasons for this failure are diverse, religious as well as scientific. To liken the earth to the planets, by making it, like them, revolve round the sun, was to be guilty of impiety, for it abolished the distinction between the corruptible matter of the earth and the incorruptible essence of the stars. The hypothesis of Aristarchus was also contradictory to the then known laws of physics, since the earth, being composed of the heaviest elements, must necessarily occupy the centre of the universe. Lastly, this hypothesis by its use of circular movements alone, did not account for the inequality of the seasons. For these reasons we can well understand why it was not followed up.

The solution of the difficulties which the system of Eudoxus could not overcome was sought in another direction. Hipparchus and Ptolemy, using the works of Apollonius, had recourse to a combination of eccentrics and epicycles. An eccentric movement is that described by a circle turning round a point within it other than its centre. A system of epicycles is formed by an arrangement of successive circles such that the centre of one is at a point on the circumference of the other. It is therefore necessary first to observe the stationary points, the retrogradations and the variable brightness of a planet, and notice the differences according to the region of the heaven it traverses, and then find the combination of epicycles

ASTRONOMY 175

and eccentrics which will account for the facts observed.

Hipparchus acquitted himself of this task in a masterly fashion. He not only succeeded in surmounting the difficulties which had arrested his predecessors, but he discovered new facts such as the precession of the equinoxes and gave a geometrical explanation of them.[1]

Inspired by the conceptions of Hipparchus, Ptolemy summarized and completed the astronomical knowledge of antiquity in the form in which it was bequeathed to the Middle Ages. At this period, two tendencies manifested themselves: one amongst the Arabs, the other amongst scholastic thinkers.

The Arabs could not be satisfied with the abstract conceptions of Greek astronomers; they sought undauntedly to materialize the geometrical fictions, and to give them a physical basis. " In reality," said Averroës, " the astronomy of our time does not exist; it is suitable for calculation, but does not agree with what really is." [2] To fill this gap Al-Bitrogi imagined nine solid and transparent spheres and attempted to explain all the celestial phenomena by their arrangement.[3] This realistic conception found favour in the Middle Ages. As Paradise was situated at the outermost part of the heavens, in order to reach it it was necessary to cross the solid spheres by certain fixed

[1] He ignored the physical cause of this phenomenon, namely the equatorial bulging of the earth. In consequence of this bulging, the earth in its movement of rotation moves like an oscillating spinning-top, therefore the plane of the equator and the plane of the ecliptic do not intersect at the same point at the end of an annual revolution. The result is that after each year the sun returns to the equinox slightly sooner than it otherwise would do with respect to a star taken as a guiding mark of reference.

[2] 13 Duhem, *Système* II, p. 139.
[3] *Ibid.*, p. 149.

paths. The journey, under these conditions, was not easy, as is shown by the fabliau " of the villein who gained Paradise by pleading ":

> *A son chevet par grand hasard*
> *Il ne se trouva pas un diable, pas un ange*
> *Qui pût le réclamer au moment du départ.*
> *Embarrassé le pauvre hère*
> *Partit sans guide et ne sachant que faire.*
> *Par bonheur il rencontre et suit l'ange Michel*
> *Qui menait lors un bienheureux au ciel.*

The scholastic philosophers, particularly Thomas Aquinas, preserved the attitude adopted by the Greek astronomers, whose hypotheses they discussed very freely. "It might be possible," declared Thomas Aquinas, " to explain the apparent movements of the stars by some other method not yet conceived by man." [1]

We know how Copernicus during the Renaissance brought into fame the heliocentric system proposed by Aristarchus, while at the same time, like the latter, he kept the conception of a finite universe. Under these conditions his hypothesis could not have a revolutionary character. Being regarded as a mathematical speculation, it was studied from this point of view and was found wanting, even by thinkers such as Tycho Brahe. It contradicted the physics of Aristotle without supplying the proofs required; moreover it scarcely simplified the calculations at all, since the movement of the planet Venus, for instance, still required a machinery of five epicycles.[2]

In order to disturb beneficially the minds of men and to find credence, the hypothesis of Copernicus needed to be completed:

1. By the considerations of Giordano Bruno on the

[1] 13 Duhem, *Système* III, p. 354.
[2] 24 Sageret, *Système*, p. 194.

ASTRONOMY

relative movements and the infinite magnitude of the universe ;

2. By the hypothesis of Kepler regarding the elliptic movement of the planets, a brilliant hypothesis, since it led to a very great simplification in the calculations, without being contrary to the appearances ;

3. By the researches of Galileo on weight, and by his observations on sun-spots ; for the results thus obtained finally demolished the physical theories of Aristotle concerning loci and the opposition between the celestial and sublunar regions.

It was therefore owing to the works of Kepler and Galileo that the mathematical and physical hypotheses could harmoniously blend and that astronomy could enter upon new paths.

CHAPTER III

MECHANICS AND PHYSICS

TO build up, as did the Greeks, a scientific astronomy which was altogether different from astrology, is a task which presents very great difficulties; but when it is a question of explaining physical and mechanical phenomena, these difficulties become almost insurmountable. In this domain we come into collision with such a variety of aspects that it seems impossible to derive them all from a small number of primary notions.

A badly-hewn tree trunk is in equilibrium on a beam. We feel instinctively that the equal division of the weight round the point of support is the cause of this phenomenon. But how can it be explained accurately? And is the equality of weights the sole cause? A bag of sand placed on a bar of iron can remain in equilibrium even if the sand is not equally distributed on the two sides of the bar.

A piece of deal and a piece of cork of the same size are thrown into the water. The latter sinks less than the former. Is it possible to explain this fact by means of the same theories which make comprehensible the state of equilibrium of the beam or of the bag of sand?

Again, it is quite another matter if we pass from the study of bodies at rest or in equilibrium to the study of bodies in motion. We know that a stone falling freely from the height of a tower accelerates its fall. How is this increase of speed to be accurately measured?

MECHANICS AND PHYSICS

We know also that a pebble thrown almost vertically by means of a sling stops at the highest point of its path and then falls back again. But what path has this pebble travelled and what has been its speed at each instant? Can we hope to deduce the explanation of such diverse phenomena from a few conceptions and a few principles?

It must be stated at the outset that the Greeks did not succeed in realizing this ideal or at least they could only do so imperfectly. It is the opinion of many thinkers that the Greek mind was too logical to be able to create sciences exclusively based on experience and experimentation. The reproach stated in these terms is certainly unjust. The Greeks were able not only to observe but to control phenomena as far as they were in a position to do so with the instruments at their disposal. G. Milhaud has clearly brought out this point, which proves the truth of the technical inventions of the Greeks and of the physical concepts which guided them.[1]

1. TECHNICAL INVENTIONS AND PHYSICAL CONCEPTS

We already find in Homeric times an advanced technique, especially in the construction of swing-doors and their fastenings (*Odyssey*, xxi, 42).[2] A little later, at the time of Thales, the engineer Eupalinus constructed in the island of Samos a tunnel which passed under the hill of Kastro. This was dug out from the two sides of the hill at the same time and the meeting-point of the miners was almost exact, which implies quite advanced methods of triangulation. In Magna Græcia in the south of Italy, Archytas, the disciple of Pythagoras, became celebrated for his mechanical inventions and discovered the use of the

[1] 21 Milhaud, *Etudes*, p. 257.
[2] 10 Diels, *Antike*, p. 34.

pulley (Aulus Gellius, X, 12). However, it was especially engines of war which appeared at the court of Dionysius the Elder towards the year 400 B.C., to be developed a century and a half later by the genius of Archimedes.

Besides powerful cross-bows and formidable catapults stretched by means of a windlass, the Greeks had even conceived the idea of the machine-gun : an ingenious mechanism made balls of metal slide automatically in the groove of a cross-bow each time it was drawn.[1]

The works of Hero show us also that the Greeks already knew how to utilize currents of hot air, and compressed air, and that they were on the way to discover the motive power of steam, as is shown by the æolipile. This apparatus is composed of a hollow sphere pivoted horizontally, which is supplied with steam from a boiler through one of the pipes serving as a pivot. This steam escapes from the sphere in opposite directions by two pipes situated at the opposite ends of a diameter perpendicular to the axis of rotation. By this arrangement the escape of the steam causes the sphere to revolve with increasing rapidity (Hero, I *Pneumatica*, p. 230). In these works there is a description of a lift and force pump for use in case of fire (Hero, I *Pneumatica*, p. 133), and also the description of a hodometer similar in all points to our taximeter. A small pin is fixed to the hub of the carriage wheel, at each turn it moves a horizontal wheel with spaced teeth. An ingenious system of toothed wheels and endless screws transmits the movement and turns the hands of the meters which mark units of different magnitudes (Hero, III, *Rationes dimetiendi*, p. 292).

The construction of the automata employed in the temples and theatres likewise reveals an intelligent use of the physical forces then known. A mechanism

[1] 10 Diels, *Antike*, p. 93.

was cleverly hidden underground just beneath the altars and communicating with them. Currents of hot and cold air, or streams of hot and cold water, or sometimes compressed air, could be used at will. All that was necessary was to light the fire on the altar. This fire heated the air and the water which worked the subterranean mechanism. This in turn acted on the statues, doves, etc., which the people then beheld moving mysteriously. The gods and goddesses raised their arms to bless the crowd and shed tears or poured out libations. Or again a dove, lifted by the hot air, rose by itself and fell to the ground (Hero, I, *Pneumatica*, p. 338 *et seq.*). It is needless to dwell on this point; the interest to us of these constructions is the degree of physical and mechanical knowledge which they imply.

In this respect the forces recognized by the Ancients in the realm of physics were fire, air and gravitation, and also magnetic force.

Plato spoke of the stone which Euripides called Magnetic, and which was generally called the stone of Hercules, which not only attracted iron rings but imparted to them its own virtue (*Ion*, 533 D). He attributed this attraction to the following phenomenon: a fluid exudes from the pores of the magnet or of the amber rubbed, and as a vacuum cannot exist in nature, the air rushes into the pores and its movement draws objects towards the magnet or electrified body.

As regards the air, the Ancients knew that it tended to rise or descend according as it is heated or cooled, and that, when compressed, it escapes with violence. They also knew that if the air be sucked up from a tube half plunged in water, the water rises in the tube, and they explained the fact as follows: bodies are superposed in order of density, at the bottom the solids and liquids, above them the air, then the fire; they always tend to follow one another in this order without leaving any space between. Moreover the force of attraction is not at all the same between all these elements. It is little felt between a liquid and a solid, but it is felt much more between a liquid and the air. This is why the air sucked up out of a tube half plunged in water attracts the water strongly and counterbalances its weight. There is equilibrium when the weight of the column of water raised is equal to the force of the attraction of the air.

The Ancients also admitted that sound is propagated in the air by spherical waves (Vitruvius, *de architect.*, Bk. V), and that it can be sent back by an obstacle and produce an echo.[1]

They admitted as well that light is propagated in a straight line, and that it is reflected on a polished surface at an angle equal to the angle of incidence. This law seems to have been known by Plato, judging by certain passages in the *Timaeus* (45 B and particularly 46 B); it was clearly enunciated by Euclid, who demonstrated its principal consequences (Euclid, VII, *Optica*). Refraction was also studied, chiefly by Ptolemy.[2]

The property possessed by concave mirrors of giving an enlarged image of an object was certainly utilized. The Ancients were also acquainted with magnifying lenses, although they did not know how to combine them for the construction of telescopes or binoculars or even eye-glasses. In the *Clouds* of Aristophanes (Act II, Scene 1) Strepsiades undertakes to efface by means of a lens the characters engraved on a tablet of wax : " When the registrar has written his summons against me, I shall take the glass and standing thus in the sun, I shall make his writing melt." Seneca,

---

[1] A. de Rochas, *La Science des philosophes et l'art des thaumaturges dans l'antiquité*, Dorbon Ainé, Paris, pp. 35 and 39.
[2] On the beginnings of mathematical physics, see 17 Loria, *Scienze esatte*, p. 557.

in his *Quæstiones Naturales* (Bk. I, Ch. vi., 5), says that small letters looked at through a glass ball full of water appear magnified. In the realm of mechanics the Ancients knew that a movement can be transmitted by means of toothed wheels and endless screws, and that it is possible to produce great effects with a small force, by allowing it time and by using a system of pulleys in sufficient number; they also knew that water is incompressible and that this property can be utilized.

Thus the technique of the Greeks was highly developed, and was well on the way towards the discoveries which came to light during and after the Renaissance. If its efforts failed to obtain greater results, it was probably because the cheapness of slave labour rendered the construction of machines unnecessary.[1] Leaving aside this important question, it remains to be seen if and how the technical results obtained were interpreted from a theoretical point of view. With the exception of some passages from Plato, it was Aristotle who first attempted to formulate in order the general laws of physics and mechanics.[2]

2. ARISTOTELIAN DYNAMICS [3]

Strictly speaking, Aristotle does not distinguish, as do modern scientists, between statics and dynamics; he does not separate the theory of equilibrium from

[1] E. Meyerson, *Bulletin de la Société française de Philosophie*, Feb.–March, 1914, p. 103.

[2] On the conceptions anterior to Aristotle, see *Evolutionnisme et platonisme*, by R. Berthelot, p. 139, the chapter entitled: *L'idée de physique mathématique et l'idée de physique évolutionniste chez les philosophes grecs entre Pythagore et Platon*.

[3] For the general characters of Aristotelian physics, consult A. Mansion, *Introduction à la physique aristotélicienne*, Louvain, 1913; and H. Carteron, *La notion de force dans le système d'Aristote*, 1924.

the theory of motion ; he does not assign to the former its own principles quite independent of the latter ; he deals generally with the movements which can take place in a mechanism ; when no movement takes place the mechanism remains in equilibrium.[1] It must not be forgotten, moreover, that, for Aristotle, mechanics as a whole rested mainly on philosophical doctrines regarding the nature of movement and of natural position, the distinction between celestial and sublunar bodies, the opposition of natural and " violent " movements, etc.

The idea of motion had primarily a much wider meaning than that which we give it.[2] As a matter of fact, by motion Aristotle understood :

1. A *substantial* change, which, for a given body, can take place in two opposite senses : the passage from form to formlessness which causes corruption, or, inversely, a passage from formlessness to form which gives rise to a birth.

2. A quantitative change, owing to which a body is diminished or increased in volume.

3. A qualitative change which causes in a body a transformation of its properties.

4. A *local* movement which brings about the displacement of a body from one position to another.

Of these four species of motion, the qualitative change presents a special character because it cannot be reduced to a mechanism or to a simple study of spatial ratios. A substance which changes in quality does so, not by a displacement of its molecules, but by an internal variation of its nature. The changes in quantity and substance, on the contrary, imply a local movement. This latter is therefore the most important in mechanics. Besides, it concerns the incorruptible celestial bodies as well as the terrestrial

[1] 11 Duhem, *Origines*, I, p. 5.
[2] 13 Duhem, *Système*, I, p. 161.

MECHANICS AND PHYSICS

bodies which are subject to the phenomena of birth and death.

This being so, the local movement can assume two forms, the one natural, the other violent.

The natural movement arises from the fact that for each body there is a place in which it exists in perfect equilibrium and towards which it naturally tends. This natural movement is necessarily simple like each of the simple substances affected by it.

Only two kinds of simple movements exist, the movement of rotation, which Aristotle calls the circular, and the movement of translation, which he called the rectilinear [1] (*Phys.* 261 b). The circular movement is that which belongs by its nature to celestial bodies, for it is, like them, perfect. The rectilinear movement, on the contrary, is the movement of bodies situated in the sublunar regions, which are subject to generation and corruption.

The simple movements of translation are of two kinds, some are directed towards the centre of the universe, others follow directions issuing from this point; the rectilinear centripetal movement (downward movement) naturally affects the heavy or weighty bodies whose position of equilibrium is the centre of the universe; the rectilinear centrifugal movement (upward movement) belongs to the light bodies which are situated in the concavity of the lunar orbit. Of the four elements which exist in the sublunar region, two are heavy, namely earth and water, and two are light, air and fire.

Thus heaviness and lightness impart rectilinear movements to the bodies possessing these qualities; but these movements cease as soon as the bodies have reached their position of equilibrium, that is to say the region of space in which they are naturally in equilibrium. So this position is not only a reality

[1] 13 Duhem, *Système*, I, p. 205.

but it possesses a certain power (*Phys.*, 208 b, 10). This fact explains why the fall of heavy bodies is accelerated; the force of the weight increases in proportion as the body approaches its position of equilibrium.[1]

The movements enumerated above, rectilinear downwards for heavy bodies, rectilinear upwards for light bodies, circular for celestial bodies, are, as natural movements, in opposition to violent movements, which result from an external constraint and which are not directed towards the position of equilibrium of a body; such, for instance, as the throwing of a projectile and the towing of a vessel.

Further, whether the movement be natural or violent, it can only be either rectilinear or circular or composed of both, " for all that which is in motion is moved either circularly or rectilinearly or both " (*Phys.*, 261 b, 25).

In postulating this principle Aristotle foresees one of the most fruitful theorems of modern kinematics which may be formulated thus: in its most general form, an infinitely small movement of a solid body is composed of an infinitely small rotation around a certain axis and of an infinitely small translation parallel to this axis.[2] However, by applying this principle without any consideration of the infinitesimal, the Aristotelian dynamics was bound to lead to manifest errors. Consider, for example, a stone which, thrown into the air by means of a sling, falls back to the ground. To the disciples of Aristotle, the trajectory described by the stone is not a parabola, but it is composed of two straight lines which are joined by a circular arc.

Having once established the distinctions between

[1] 16 Jouguet, *Lectures de mécanique*, I, p. 3.
[2] 13 Duhem, *Système*, I, p. 171.—24 Sageret, *Système*, p. 214.

MECHANICS AND PHYSICS

celestial and terrestrial (light and heavy) bodies, and between natural and violent movements and their kinds, Aristotle defined the conditions and laws of all motion.

In his eyes any body which is moved is necessarily subjected to two influences, a force and a resistance; without the force it would not be able to move, but without the resistance, its movement would be accomplished in an instant, and it would immediately reach the point to which it is impelled by the force; the velocity with which a body moves depends therefore both on the magnitude of the force and the magnitude of the resistance.[1]

This being so, if bodies of different weights, balls, for instance, of the same material and of various sizes, are placed on a plane horizontal surface, and if each of them is pushed at the same time with the same force, the lighter balls will roll more quickly and further than the heavy ones. From this, and other analogous facts, Aristotle deduces the following law which he considers the basis of mechanics.

The force F which moves a body is equal to the resistance R which acts on this body, multiplied by the velocity V imparted to it by the force

$$F = RV.$$

This law of mechanics excludes the possibility of empty space in nature, for if empty space existed anywhere, bodies would not be subject to any resistance when passing through it, and the ratio F/R which expresses the velocity would lose all numerical significance (*Phys.*, 216 b). Thus, the existence of empty space is far from being that which rendered movement possible, as the atomists pretended; on the contrary it is inconceivable that a body may move in empty space with a local movement.[2]

[1] 13 Duhem, *Système*, I, p. 192.   [2] *Ibid.*, p. 197.

Further, according as the movement is natural or violent, the resistance and the force manifest themselves differently. In the natural movement, the force is constituted by the quality of heaviness or lightness which impels a body towards its position of equilibrium and which acts inexhaustibly until this point is reached by the body. As to the resistance, it is simply that offered by the medium traversed, for instance the air in the fall of a heavy body.

Observation shows us, besides, that the natural movement, in as far as it is rectilinear, is accelerated (*Simplicius in Aristotelis.* Diels, Bk. V, ch. vi, p. 916).

When a streamlet of water falls from a height, from a gutter, for example, it appears continuous near its origin, but soon the acceleration of the fall detaches the drops of water from one another and they fall to the ground separately.

When a stone falls from a height, it strikes an obstacle more violently if it is stopped towards the end of its fall than at the middle or beginning; this more violent impact is the sign of a greater velocity.[1] Moreover, the theory confirms the observation. The rectilinear movement cannot go on for ever, it has a beginning and an end. Hence, starting from rest at a determinate moment of the duration, a moving body only passes from a zero velocity to a given velocity by means of an acceleration, and this acceleration continues for the same reasons as it began. It only ends when the moving body has reached its goal, its position of equilibrium.[2]

In violent movements such as the traction of a cart and the towing of a vessel, resistance is represented by the weight of the object to be moved, and force by the motive power continuously acting on this object. The movement of a projectile in the air is a special

[1] 13 Duhem, *Système*, I, p. 388.
[2] 24 Sageret, *Système*, p. 214.

MECHANICS AND PHYSICS 189

case. Here, it is the air which plays the part of motive power. When displaced by the projectile coming out of the catapult or sling, the air flows back behind the projectile and pushes it forward. Whilst the rectilinear natural movement is accelerated, the violent movement is of necessity retarded (*Phys.*, 230 b, 25).

From the mechanical point of view, the interest of Aristotle's teaching lies in the law of proportions which he establishes, as we have seen, between the velocity V, the force F and the resistance R. The same force can move successively a heavy body and a light body; but it will move the heavy body slowly and the light body quickly; thus the velocities of the movements imparted to these bodies will be inversely proportional to their weights. "The velocity of the lighter body will be to the velocity of the heavier body as the weight of the heavier body is to the weight of the lighter body" (*De Coelo*, 301 b).

This law appears to be a faithful translation of common observation. At first sight, it even seems to apply to the free fall of bodies in space. In this case the motive force is the weight, the resistance is the air. As a matter of fact, a light body like a feather falls more slowly than a heavy body like a piece of lead. If, however, we take two bodies of the same shape but weighing respectively 1 lb. and 2 lbs., we ought to have, since the resistance of the air is the same,

$$1 \text{ lb.} = RV \text{ and } 2 \text{ lbs.} = R_2V.$$

The body weighing 2 lbs. should fall twice as quickly as the one of 1 lb., which is contradicted by experience.

Thus the law postulated by Aristotle, which persisted until the Renaissance, is manifestly false. The resistance of the air does not play the part attributed to it by the Stagirite, and bodies fall with equal speed in empty space as had been supposed by the Atomistic philosophers and with them Lucretius (II, 235): "Consequently the atoms, in spite of the inequality of their masses, must move with equal velocity in empty space."

Again let us take a body subject to a force which remains the same and to a resistance which continuously increases until it becomes equal to the force, for example, when a stake is driven into the sand. Experience teaches us that the velocity becomes nil at a given moment, but, according to the law of Aristotle, that is impossible, since we have the constant:

$$V = \frac{F}{R}$$

Aristotle saw this difficulty, but in order to remove it he simply laid down the law that a small force cannot move a large body. "Because a whole force moves a body along a certain distance, it does not result that half this force moves this body along any distance during any time. A single man would in that case be able to move the ship which all the haulers pull, if, the force of the haulers being divided by a certain number, the distance traversed were also divided by the same number."[1] Aristotle could not explain by his theory why it is easier to move with a given force a carriage having large wheels than one having small wheels. His mistake lay in considering as simple and elementary, facts which are really very complex.

From the formula he had stated, $F = R \times V$, Aristotle drew the conclusion that the properties of the lever and the balance are related to the study of the velocities with which circular arcs are described. Two forces are equivalent if by moving unequal weights with unequal velocities they give the same value to the product of the weight by the velocity.

[1] *Phys.* 250 a, 10, quoted from 13 Duhem, *Système*, I, p. 194.

MECHANICS AND PHYSICS

"If we take a rectilinear lever divided by a fulcrum into two unequal arms to the ends of which two unequal masses hang; when the lever turns round its fulcrum, the two weights will move with different velocities, the one which is farthest from the fulcrum will describe in a given time a greater arc than the one which is nearest to it; the velocities with which the two weights move have the same ratio to each other as the lengths of the arms of the lever.

When, therefore, we wish to compare the forces of

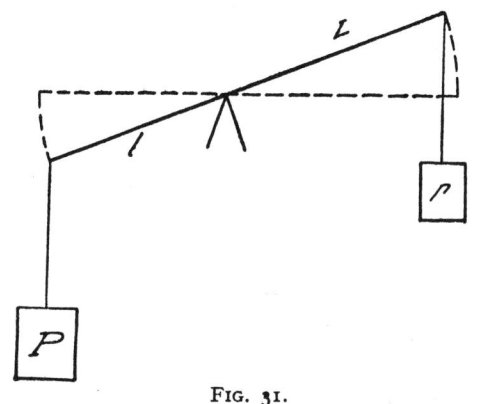

FIG. 31.

the two weights, we must find, for each of them, the product of the weight by the length of the arm of the lever; that one which corresponds to the greater product will outweigh the other; and if the two products are equal, the two weights will remain in equilibrium." [1]

By an intuition of genius, Aristotle extended to other mechanisms his theory of the lever; he shows that the various operations of these mechanisms can be explained merely by considering the velocities with which certain circular arcs are described; hence he

[1] II Duhem, *Origines*, I, p. 7.

foreshadows the principle of virtual velocities. "For," said he, "the properties of the balance are reduced to those of the circle; the properties of the lever to those of the balance; and the greater part of the other peculiarities of mechanical movements are reduced to the properties of the lever" (*Quæstiones mechanicæ*, 848 a, 11).

Aristotle, however, was not able to deduce from the principle which he discovered all the rigorous consequences which arise from it. He applies it to problems which are too complex for the means by which he attempts to solve them. Already as regards the lever he had been confronted with the following difficulty: "the line described in a movement of the lever through the point of application of the force of resistance is a circumference of a circle; it does not coincide with the vertical line along which this force or resistance acts."[1] Aristotle perceived the problem, but he did not succeed in solving it. He contented himself with supposing that a balance is more accurate the longer its arms are, for then the circular arc described approximates more nearly to a vertical line.[2]

3. ARCHIMEDES AND STATICS

The method adopted by Archimedes is very different from that of Aristotle. Archimedes limited the domain of theoretical mechanics to the study of problems of equilibrium, and in this manner he succeeded in establishing the foundations of statics and hydrostatics. He did not dream of seeking his fundamental hypotheses in kinematics, for the laws which govern the displacement of bodies in space do not seem to be reducible to the intelligible and clear conceptions of reason. On the other hand, the phenomena of equilibrium can be interpreted by means of very simple rules,

[1] 11 Duhem, *Origines*, I, p. 9.
[2] 16 Jouguet, *Lectures de mécanique*, I, p. 35.

by a method similar in all points to that adopted by Euclid in his *Elements*.

This being so, Archimedes only required that the two following propositions should be granted :
 1. Two equal weights applied at equal distances from the fulcrum are in equilibrium.
 2. Two equal weights applied at unequal distances (from the fulcrum) are not in equilibrium and the more distant weight descends.

These and similar postulates were considered by Archimedes to be self-evident and independent of all experience. If a rod supposed to have no weight rests freely at its middle point on a fulcrum, and if

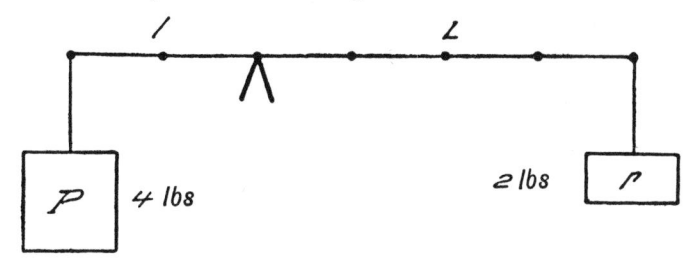

FIG. 32.

two equal weights are suspended from its extremities, it would appear *a priori* that the whole system is in equilibrium, for as the system is symmetrical there seems no reason why a movement should take place in one direction more than in another. It seems therefore evident, by virtue of the law of adequate reason, that the hypothesis is independent of all experience.[1]

When this hypothesis is admitted, the law of equilibrium is easily established in the case of a lever, namely $pL = Pl$, the relation in which the greater force P is exerted at the shorter arm l of the lever (Fig. 32).

[1] Mach, *La Mécanique*, Hermann, Paris, 1904, p. 18.

To demonstrate this relation it is sufficient to replace, in the example given, the weight of 4 lbs. by an arrangement of two weights of 2 lbs. each, then there will be symmetry round the fulcrum and consequently equilibrium (Fig. 33).

After having established the law of the lever, Archimedes used it in the investigation of the centre of gravity of various surfaces such as triangles, trapeziums, and segments of a parabola. He demonstrated, for instance, that the centre of gravity of a triangle is the point of intersection of the medians. In fact, if

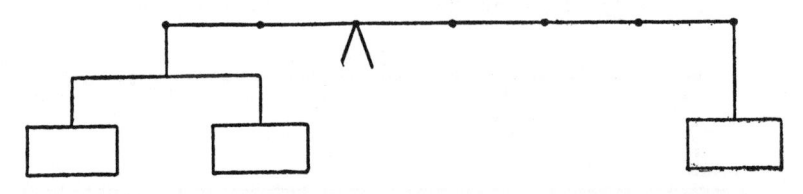

FIG. 33

a triangle be placed on the blade of a knife in such a manner that the latter coincides in each position with one of the medians, the triangle is in equilibrium. Consequently, it will also be in equilibrium if it be suspended by the point of intersection of the medians.

By a similar method, but making use of new hypotheses, Archimedes demonstrates in a masterly fashion a series of propositions in hydrostatics, which are still renowned. Amongst other things, he proves that a body plunged in a fluid of equal density to its own is entirely immersed, but remains suspended in the fluid ; and that a solid floating in equilibrium on the surface of a liquid displaces a weight of this liquid equal to its own weight. It can be clearly seen that, in mechanics, Archimedes did not, like Aristotle, deduce his principles from the general laws of motion. He based his theories on certain simple laws of equilibrium, taken as self-evident. Thus he made the science of equilibrium an independent science which owes nothing to the other branches of physics; he established statics.[1] But the rigour and lucidity which he obtained were bought at the price of a real sacrifice of the generality and fecundity of the method.

The laws which govern the equilibrium of two heavy bodies suspended from the arms of a lever have been deduced from hypotheses peculiar to this problem. They are of no use when there arises a case of equilibrium in entirely different conditions; when analysed, they cannot give any indication as to the choice of new hypotheses. So that, when Archimedes studied the equilibrium of floating bodies, he was obliged to have recourse to principles which have no analogy to the requirements of the theory of the lever.

As M. Duhem remarks: "Although an admirable method of demonstration, the path followed by Archimedes in mechanics is not a method of invention; the certainty and lucidity of his principles are largely due to the fact that they are gathered, so to speak, from the surface of phenomena and not dug out from the depths."[2]

It seems to us that this is the reason why the demonstration of Archimedes is not entirely satisfactory even from a logical point of view, for it finally comes back to the disguised verification of a fact.

Doubtless, by virtue of the principle of symmetry, we can logically maintain that two equal weights A and B suspended from two equal arms of a lever will be in equilibrium; but we cannot know *a priori* what will happen if we replace one of these weights A by two smaller weights a and a_2 which are in equilibrium and whose sum is equal to A. Experience

[1] 11 Duhem, *Origines*, I, p. 11.
[2] *Ibid.*, p. 12.

alone can inform us on this point. An example will show more clearly that this is so.[1] Let us consider a compound pendulum formed of a rigid rod of negligible weight to which are fixed a weight of 2 lbs. at a distance of 4 inches, and another weight of 2 lbs. at a distance of 8 inches (Fig. 34).

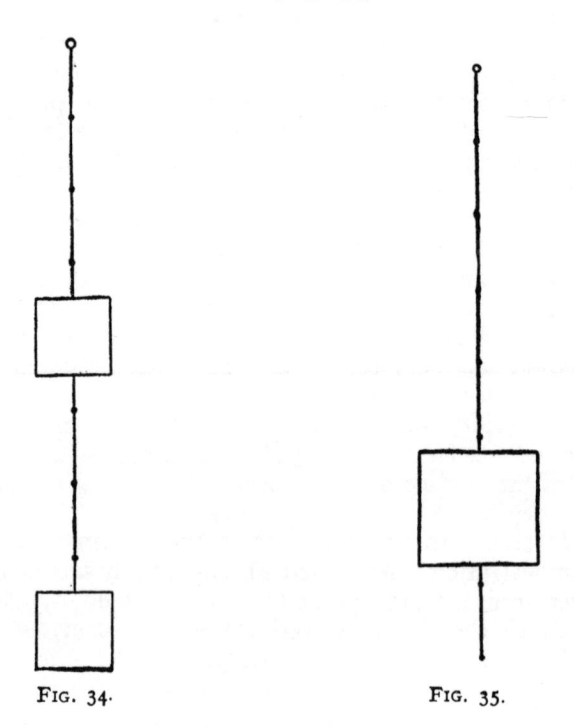

FIG. 34.   FIG. 35.

When the pendulum is held in a horizontal position the moment of the force acting on it is equal to
$$2 \times 4 + 2 \times 8 = 24.$$
According to the reasoning of Archimedes, we can

[1] Cf. L. Lecornu, *La Mécanique*, Flammarion, Paris, 1918, p. 56.

replace the two weights by a single weight equal to 4 lbs. and fixed at a distance of 6 inches. The moment of the force acting on the pendulum in a horizontal position is still equal to 24, i.e., the product of 6 by 4.

Under these conditions it would seem that if we allow the pendulum to oscillate, we must obtain the same result in both cases and find that the duration of the oscillations is the same. But in fact it is not so. Why? Because the conditions of symmetry for a system in motion are not the same as for a system in equilibrium. By changing the compound pendulum into a simple pendulum we certainly have not changed the static moment of the system but we have modified its moment of inertia, and for this reason the times of oscillations can no longer be equal.[1]

Thus from the logical principle of symmetry one cannot *a priori* deduce consequences before making any experiment. It is experience alone which can teach us in what way this principle works in nature, for a mass of unknown factors may interfere and confuse its application just where the latter might rightly be expected. Concerning the lever, we know for example that to maintain equilibrium, it is immaterial to hang the arrangement of two weights higher or lower than the weight it replaces, and to place this arrangement parallel or perpendicular to the direction of the lever.

If, notwithstanding, the demonstration of Archi-

---

[1] The moment of inertia I is the sum of the masses m multiplied by the squares of their distances r from the axis of suspension. The time of oscillation is then equal to $T = 2\pi\sqrt{\frac{I}{M}}$ when M represents the static moment.

If the calculations in the chosen numerical example be made, it will be found that I for the compound pendulum is equal to $2 \times 4^2 + 2 \times 8^2 = 160$, while for the simple pendulum it is only equal to $4 \times 6^2 = 144$.

medes is accurate, it is solely because it is based on an intuitive empirical statement, namely that the effective power of a force at a given moment is equal to this force multiplied by its distance from the vertical axis which passes through the fulcrum $Pd = pD$ (Fig. 36).

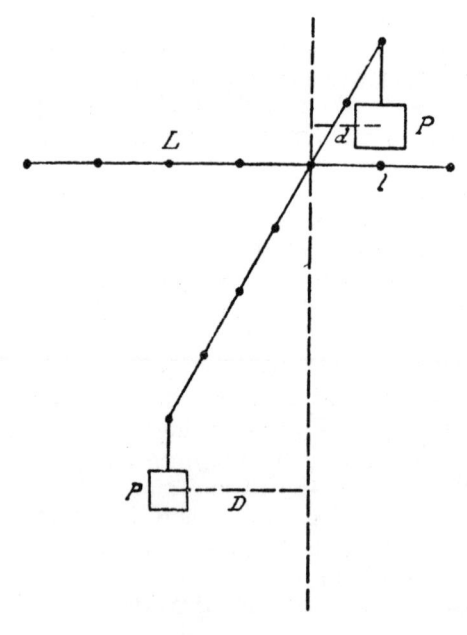

FIG. 36.

This relation, which is based on the moments of forces, was not clearly formulated until the end of the Middle Ages; it is equivalent, in a horizontal position, to the relation $Pl = pL$; it was by this that Archimedes was instinctively guided.

As we have seen, from a desire for lucidity he confined his theoretical researches to statics, that is, to

MECHANICS AND PHYSICS

a very special class of phenomena; and for this reason his method, in the course of time, proved less fruitful than the dynamical conceptions of Aristotle.

At first sight it may seem surprising that Archimedes, after having invented and perfected so many ballistic machines, did not attempt to study their theory. This abstention may be accounted for by the logical difficulties raised by the idea of motion. The arguments of Zeno of Elea on this point had produced in the minds of the ancient philosophers an uneasiness which was never dispelled.

For example, the space in which a body moves is motionless: how are we to understand the relationship of a moving body to something motionless? Look at the arrow in flight. It follows an immovable line which is its trajectory, and it must at each instant coincide with a portion of this trajectory since it traverses it. Now it cannot do this without itself coming to rest for an instant, however short, therefore its whole movement is a sum of instants of rest.

To the Greek geometers it did not appear possible to avoid the objections raised by Zeno, and this perhaps was the reason that Archimedes did not attempt to establish the foundations of rational dynamics.

4. LATER DEVELOPMENTS

It would be a mistake to consider the works of Aristotle and Archimedes as isolated examples of their kind.[1] The statics of Archimedes, in particular, by its subtle analysis and marvellously clever solutions, the interest of which is not apparent to the uneducated, bears wituess to a science already far advanced, and in no way resemble the uncertainties of a science newly born. Moreover history confirms this supposition, since it places prior to Archimedes the mechanical

[1] 11 Duhem, *Origines*, II, p. 280 *et seq.*

problems perhaps falsely attributed by tradition to Aristotle, and which enunciate with remarkable accuracy the composition of movements by the parallelogram of forces. If this work is not Aristotle's, its distinctly peripatetic inspiration points to its being due to one of his immediate disciples.[1]

Another tradition preserved by the Arabs attributes to Euclid various treatises on the lever and heavy and light bodies. These may not have been the work of Euclid, but they were certainly written by one of his contemporaries; for, whilst drawing their inspiration from peripatetic dynamics they use an axiomatic method similar to that of the *Elements*, but much less elaborate than that of Archimedes.[2]

If Archimedes had precursors, he assuredly had also followers in antiquity. Byzantine and Alexandrian science pursued the various paths opened up by him. The art of engineering, developed by him to such a high degree, inspired, as we have seen, the labours of Ctesibius, Philo of Byzantium and Hero of Alexandria. Pappus, on the other hand, endeavoured in theory to equal the demonstrations of the great Syracusan. He alone of all the geometers of antiquity attacked the problem of the inclined plane, without, however, succeeding in solving it correctly (*Pappus*, Hultsch edit., pp. 1032 and 1033).[3] On the other hand, he discovered the two following theorems, which are known by his name, though sometimes called the theorems of Guldinus (*idem*, p. 652), namely:

The volume generated by the revolution of a surface, bounded by a curved line, about an axis is equal to the product of the area of the surface and the circumference or arc of circumference described by its centre of gravity.

The surface generated by a curve turning round an

[1] II Duhem, *Origines*, I, p. 108.   [2] *Ibid.*, p. 67.
[3] *Ibid.*, p. 144.

axis is equal to the product of the perimeter of the curve and the circumference or portion of circumference described by its centre of gravity.

The foundations which Aristotle had assigned to mechanics were criticized by Joannes Philoponus, called the Grammarian or the Christian, because he was converted from Alexandrian Neo-platonism to Christianity about the year A.D. 520. In his *Commentaries* on the five later books of Aristotle's *Physics*, Philoponus disputes Aristotle's arguments against the existence of empty space, for " if the medium were solid, it would hinder the movements of bodies, which in order to move would be obliged to divide it ; these bodies nevertheless are in motion. If the medium were empty, what is there to prevent the flight of an arrow, a stone or any other thing, as long as there is an instrument for throwing, a projectile and space ? "[1] Thus the air, far from sustaining the movement of a projectile, only hinders it.

The Arabs confined themselves to accepting and commenting on the treatises of mechanics bequeathed to them by antiquity. The Western Middle Ages were more venturesome. The fragments received from Byzantium and from Islamitic science were sufficient to awaken their attention and fertilize their intelligence. From the thirteenth century, perhaps even before, the school of Jordanes opened up paths unknown to antiquity. Jordanes of Nemora, whose real name and nationality are unknown to us, discovered the following law : If a force can raise a certain weight to a certain height, it will be able to raise a weight n times greater to a height n times smaller.

Another savant worthy of mention is he whom P. Duhem calls the forerunner of Leonardo da Vinci. We know nothing about him, except that he lived later than Jordanes and was a man of genius. Inspired by

[1] Quoted from 13 Duhem, *Système*, I, p. 383.

the law discovered by Jordanes, he was able to discover the law of equilibrium of the bent lever, and the idea of moment, and also to give to the problem of the inclined plane a solution which was rediscovered by Stevinus in the sixteenth century.

It is evident, that although the Greeks displayed much ingenuity in the domain of technical applications, they were not able, except in certain special cases, to explain physical and mechanical phenomena in conformity with their ideal of science.

CHAPTER IV

THE CHEMICAL AND NATURAL SCIENCES

1. CHEMISTRY

IN the sciences which we have hitherto considered, observation and practice have, up to a certain point, guided theory. It was not the same with chemistry, the theories of which were closely connected with metaphysics and had no great influence on the technical processes. The first gropings of chemical technique are very ancient. They seem to go back to a prehistoric epoch, to the time when metals were first used for manufacturing weapons, and when certain alloys were perceived to be advantageous. Amongst these alloys, that of tin and copper was specially important. From the most remote antiquity Egypt was an important centre of the trade in tin ; which in later times was supplied by Phœnician traders.[1] Other metals were afterwards discovered and alloyed. In Egypt, the method of treating them was preserved by tradition in the form of short and probably mysterious receipts whose secret was jealously guarded by the priests. Certain hieroglyphic signs, completed by oral instructions, were sufficient to ensure the transmission of the methods of manufacture.

As to the Greeks, the sum of their practical knowledge amounts approximately to the following : " They knew how to prepare certain salts of copper, of

[1] M. Delacre, *Histoire de la Chimie*, Gauthier-Villars, Paris, 1920, p. 16.—J. de Morgan, *L'Humanité préhistorique*, Renaissance du Livre, Paris, 1921, p. 119.

potassium and of sodium, how to render fabrics incombustible, how to treat minerals. Some substances, like alum, were used for the same purposes as at the present time. The manufacture of pigments, which implies chemical reactions, was far advanced at the time of the great Greek painters. But it was more particularly in the preparation of poisons that antiquity excelled. Owing to the limitations of the science of the time, the same name was often given to very different substances. Thus χαλκάς denoted either copper, or its various alloys with tin, zinc, lead or other metals." [1] The Romans merely practised, without developing, the science which they received from Greece and Egypt.

Although very ancient, chemistry did not produce any systematic publications until relatively late. In fact it was in the Alexandrian period, under the Ptolemies, that there appeared for the first time a work summarizing the metallurgical and chemical knowledge of the period. This work was published under the name of Democritus, but its author was in reality a certain Bolos, who lived about 250 to 200 B.C. Inspired by him, there arose a series of writings, the most important of which is entitled *Physica et Mystica* by Democritus, which, in four books, treats of gold, silver, pearls and precious stones, and lastly of the manufacture of purple. It is probable that the first alchemistic and hermetic treatises also belong to this period.[2]

Unfortunately we only possess fragments of all this literature. These, however, are sufficient to show that the idea of the transmutation of elements was already common, as also the belief that a single substance (prima materia) is the base of all material bodies.

Of the manuscripts relating to chemistry the most

[1] L. Laurand, *Les Sciences dans l'antiquité*, Picard, Paris, 1923, p. 29.—For the terminology and composition of minerals, see 1 Berthelot, *Introduction*, pp. 228–268.
[2] 10 Diels, *Antike*, p. 113.

THE CHEMICAL AND NATURAL SCIENCES 205

important by far were discovered in a tomb in Egypt, almost a century ago. One is called the Leiden papyrus, the other the Holmiensis papyrus.[1] They were written in the third century A.D., but their matter is much more ancient and is largely a reproduction of the *Physica et Mystica* mentioned above. It is probable that the possessor of these manuscripts had requested that they should be buried with him at his death in order to avoid trouble to his heirs; for Diocletian, from fear of coiners of base money, had caused all books treating of the manufacture of gold, silver and precious stones to be burnt.

The Leiden and Holmiensis papyri are of great importance, especially because they give exact and detailed receipts for the working of metals and the method of obtaining certain alloys (amongst others the asemon), and also how to manufacture imitation pearls, rubies and topazes. Magical and astrological prescriptions were added to these receipts, for metallurgical operations may be aided by propitious conjunctions of stars and planets. " A metal was assigned to each heavenly body. To the sun, gold; to the moon, silver; to Mars, iron; to Saturn, lead; to Jupiter, electrum; to Hermes, tin; to Venus, copper." [2]

It was in the fourth century A.D. that the terms Alchemy and Chemistry first made their appearance. For a long time their authorship was attributed to the astrologer Firmicus Maternus; but in reality these terms were introduced by the famous Zosimus of Panopolis, who lived at about the same time as Firmicus Maternus (A.D. 336). Zosimus derived the word Chemistry from the name of the Jewish prophet Chemes; according to Diels it is more probable that

[1] For the history of these papyri, see 1 Berthelot, *Introduction*, p. 4, *et seq.*, and 10 Diels, *Antike*, p. 118.
[2] 1 Berthelot, *Introduction*, p. 77.

Chemistry, or better Chymistry, comes from the Greek word χύμα (fusion).[1]

This same Zosimus traces the origin of chemical science back to the epoch before the Flood, when, according to the story of Genesis (ch. vi) afterwards enlarged in the Book of Enoch, the sons of God married the daughters of men. In order to seduce the latter, one of the former, the angel Asasel, revealed to them the secrets of the healing properties of plants and the beauty of artificial jewels. Hence the diabolical character of Chemistry.

The writings of Zosimus certainly contain valuable information as to the treatment and alloying of metals, the fabrication of precious stones, and even describe interesting processes of distillation; but they are cumbered with gnostic and magical ideas which persisted for centuries; and these gave to alchemy the character of an occult science feared by the unlearned, because its secrets belonged rather to demons than to God.

However, notwithstanding these mystical dreams, the researches of the Alchemists were directed by ideas of a philosophical and even scientific nature.

As we have seen, the Ionians, from the dawn of Greek philosophy, admitted that matter is one in its essence, but that it can assume various forms.

A century later, Empedocles formed the conception of two imponderable media, one endowed with the power of attraction, the other with the power of disintegration. These two media constantly acted on the four constituent elements of matter, namely, water, earth, air and fire. They ceaselessly united and separated these elements, and in this manner worlds and phenomena were evolved.

At almost the same epoch Democritus boldly postulated the existence of empty space, and established the

[1] 10 Diels, *Antike*, p. 110.

THE CHEMICAL AND NATURAL SCIENCES

foundations of the atomic theory. Bodies are composed of material atoms which differ from one another only in magnitude, shape and weight. These atoms by combining and separating produce all sensible phenomena.

Up to a certain point Plato combines the ideas of Empedocles with those of Democritus. According to him, mathematical facts constitute the intelligible basis of the world which the Demiurge desired to create; but in order that this world might become tangible and visible, it had to be brought into existence under the form of earth and fire. Moreover, as earth and fire cannot enter into direct relationship with one another, it was necessary to unite them by means of water and air in the following proportions : [1]

$$\frac{fire}{air} = \frac{air}{water} = \frac{water}{earth}$$

In order that combinations may be formed between these constituent elements of the universe, it is necessary that these should take the form of regular polyhedra ; therefore the earth-element will be a cube, the water-element an octahedron, the air-element an icosahedron, and the fire-element a tetrahedron. From the fact that there exists a fifth regular polyhedron, the dodecahedron (the faces of which are pentagonal) Plato deduced the existence of a fifth element also, namely, the ether. The ideas of Plato and particularly of Democritus resemble in many respects the conceptions of modern chemistry. They had, however, but slight influence on the development of the science because they evaded the methods of experimental verification which were in use until the end of the eighteenth century.

In this domain also, the conceptions of Aristotle, afterwards seen to be false, had an important influence. Aristotle begins by opposing matter and form. The

[1] 13 Duhem, *Système*, I, p. 30 *et seq.*

"prima materia" is neither fire, air, water nor earth; but it is capable of becoming all these elements. At the same time it assumes certain fundamental qualities which are irreducible to one another (white and black, cold and hot, etc.); the same body can successively receive these qualities. The task of physics and hence of chemistry is to determine in the first place all the irreducible forms which exist in nature, and then study the laws by which a body can successively assume all or part of these forms. Now experience teaches us that only the following properties are suitable for all bodies, namely heat and cold, dryness and humidity. These therefore are the properties which constitute the irreducible forms. By combining them in all possible ways six pairs are obtained of which two, the dry-damp and the cold-hot, must be eliminated as contradictions.

The four pairs which remain are represented by the following bodies: [1]

cold-damp	water
cold-dry	earth
damp-hot	air
dry-hot	fire

This conception of Aristotle is not well adapted to mathematical considerations, especially to a geometrical representation; but it appears to take into account the immediate facts of existence, and was therefore adopted in the Middle Ages by the Arabian philosophers.

These latter, however, were gradually led to modify the classification of Aristotle which does not take into account the exceptional importance of metals. According to them, mercury symbolizes metal and must form the basis of all metals. Sulphur constitutes another most important property, combustibility; the

[1] W. Ostwald, *L'Evolution d'une science, la Chimie*, Flammarion, Paris, 1909, p. 6.

earth represents non-metallic minerals, salt the solubility in water and solvent action on other bodies. These ideas relate to ideal elements yet to be discovered, and not to mercury, sulphur, earth and salt as known to us through our senses. The discovery of these elements would enable the transmutation of substances to be effected, that is to say the transfer of a property from one body to another. In particular it would be possible to transmute any metal into gold. Only, the transmutation must be effected in a certain order.

As, in the eyes of modern chemistry, an element has an affinity for certain known elements, so the alchemists held the opinion that, although every body can be transmuted into something else, this can only be done by following an invariable order. For example, if F represents iron and O gold, in order to transmute F into O it is necessary to give to F the property G, then by means of G the property H, and so on up to O. If one of the links be omitted, the transmutation will not take place. Hence the famous symbol of the serpent biting its tail.

This investigation of the characteristic circular order of the transfer of the properties of bodies could not reach its goal, but it had the result of perfecting metallurgy, the manufacture of glass and the remedies employed in medicine, and discovered, by means of distillation, several essences or spirits such as turpentine.

The history of chemistry is of a strange character. From the fifth century B.C. Democritus had laid its theoretical foundations. However, these were not verified until after the work of Lavoisier at the end of the eighteenth century. Until that time, practical research gave rise to conceptions which, while doubtless erroneous, seemed to be more in agreement with the data directly furnished by experience.

2. THE MEDICAL AND NATURAL SCIENCES

In the first part of this book we have shown the progressive development of medical science, and noted the remarkable discoveries which were due to it. It is sufficient here to recall briefly the spirit and methods which characterized these discoveries.

Like other primitive peoples, the Egyptians and Chaldeans considered disease either as a punishment sent by a Divinity, the work of malevolent spirits, or the consequence of spells wrought by man. In every case the agent of the disease was a spirit which entered the body and destroyed the tissues.

Therefore to obtain healing, the intervention of both the priest and the physician was necessary. The former had to appease the Divinity by sacrifices and prayers. The latter had a twofold task. He had to drive away the spirit who caused the disease, by exorcisms and incantations on the one hand, and on the other hand by drugs which were feared by the spirit and at the same time built up the tissues of the patient. The choice of these drugs was determined more often by a fantastic association of ideas than by specific experience. " The euphrasia was supposed to heal diseases of the eye because its corolla has a black mark resembling the pupil of the eye, whilst the red tint of hæmatite seemed to point it out as a means of stopping hæmorrhage. The Egyptians believed that the blood of black animals would prevent the hair from turning white, and even to-day in Styria, as formerly in India, Greece and Italy, jaundice is banished into the bodies of yellow birds." [1]

Greek medicine from the first took up a different position: The *Iliad*, speaking of the care of the wounded, makes no mention of superstitious practices. The wounds must be dressed with special balms and the warriors revived with wine, barley and cheese.

[1] 14 Gomperz, *Penseurs*, I, p. 294.

THE CHEMICAL AND NATURAL SCIENCES

Doubtless there existed in Greece, side by side with the scientific and lay medicine, a medical art practised by the priests and thaumaturgists, in which incantations played a preponderant part. But this fact did not prevent the lay medicine from following an entirely different direction. In accordance with the scientific ideal glimpsed by the Greek philosophers, it considered that all disease, including epilepsy, had its origin in a natural cause. The primary consideration was therefore to know the exact structure of the human body, and it was to this that Greek anatomy applied itself with conspicuous success, especially during the Alexandrian period. In the study and the treatment of diseases, Greek medicine displayed a no less remarkable skill. It held that the health of the body consisted in a state of equilibrium maintained by food and exercise. " The fundamental condition of health is to observe a just proportion between work and food, by taking into account the constitution of the individual, differences of age, season, climate, etc. A man would be protected from all disease if one of these factors—the individual constitution—could be ascertained beforehand by the doctor."[1] We have seen how Hippocrates tried by means of his humoral theory to define the conditions of right and wrong proportions which constitute health and sickness.

Whatever explanations might be suggested, Greek medicine was as a rule distrustful of philosophic opinions which could not be directly verified by experience. It only accepted hypotheses which were founded on and verified by facts. It had a very clear perception of the individual and general characteristics of diseases. Hence it succeeded in noting the symptoms and courses of most of them with remarkable accuracy, and in ascertaining causes as well as remedies. Surgical art was likewise systematically practised, and brought

[1] 14 Gomperz, *Penseurs*, I, p. 304.

to a high degree of perfection thanks to a comprehensive set of implements, as is shown by the surgical instruments discovered at Pompeii.[1]

The credit of having established the scientific bases of the natural sciences belongs to Aristotle and his disciples. It has been mentioned that Aristotle rescued zoology from oblivion. This, however, as Gomperz points out, is to honour him both too much and too little, for it is attributing to him an almost superhuman work and at the same time a mass of errors for which he is not responsible.[2] Aristotle had predecessors amongst the philosophers and especially amongst the physicians, whose opinions he often quotes either with approval or disapproval. However, although he profited by the work of his predecessors, he made more use of the observations he himself was able to make, and the information he methodically gleaned.

In the three great works which he published (*Historia animalium*; *De partibus animalium*; *De generatione animalium*) he interprets the facts observed according to finalistic views, and by considering mechanical causes as aids to final causes. According to him, the life of nature is divided into two spheres, in one of which necessity reigns, whilst the other is ruled by tendencies and by finality (*De generatione animalium*, 759 b).[2] Life is motion. Now all motion implies both a form which moves and a matter which is moved. The form is the soul, the matter is the body. The soul is the permanent force which moves the body and determines its structure. But as form only gradually overcomes the resistance of matter, the psychic life comprises three degrees: nutrition, sensa-

[1] 10 Diels, *Antike*, p. 23.
[2] 14 Gomperz, *Penseurs*, III, p. 150.—F. Houssay, *Nature et Sciences naturelles*, Flammarion, Paris, p. 62.—22a Robin, *La Pensée grecque*, p. 351 *et seq*.

THE CHEMICAL AND NATURAL SCIENCES

tion, and intelligence.[1] Having established these foundations, Aristotle explains the anatomical structure and constitution of living beings, in conformity with his doctrine, by final causes. This kind of explanation presents difficulties which Aristotle was not always able to avoid. Thus he attributed baldness to the coldness of the brain, and the timidity of certain animals to the size of their hearts. But as a rule teleological principles led him to happy results.

In his classification of animals, Aristotle had the great merit of abandoning the dichotomous division praised by Plato, which was based solely on the presence or absence of some particular feature (winged and wingless, for example).[2] According to this method of division a species is composed *a priori* of two sub-species, which in their turn are each divided into two, and so on. A classification of this kind is not organic, because it forcibly separates beings which are in reality closely allied, for instance, the winged ants (males and females) from the wingless (workers), the male fire-fly which has wings from the female which has none.

Aristotle also considered that, in classification, anatomical characteristics should outweigh physiological characteristics, which depend on the mode of existence and on adaptation. He excelled in discovering organic correlations and reciprocal dependences. He showed how the removal of a small organ can bring about a change in the whole body; how, for example, in eunuchs there is a transition from the masculine to the feminine. He formulated the law of the balance of organs. " Everywhere nature restores to one part what she takes away from another. . . . She cannot make the same expenditure in two directions. . . . It is impossible for her to expend the same material in

[1] E. Boutroux, *Études d'histoire de la philosophie*, Alcan, Paris, 1897, p. 155.
[2] 14 Gomperz, *Penseurs*, III, p. 163.

several places at the same time" (*De gener. anim.*, 750–3).[1] Aristotle also affirmed the subordination and the hierarchy of beings in the animal scale. The organic individuality becomes stronger as we pass from inferior to superior beings. Only, to Aristotle, this hierarchy was not the result of a progressive and continuous evolution, as Lamarck and Darwin were to maintain. It remains the same for all time, since different species, even those most akin, cannot form a fertile and lasting union.

Connected with the anatomical generalizations there are physiological generalizations of which Alcmaeon, Empedocles, and the followers of Hippocrates had already set the example. In this domain, Aristotle very clearly established the modern distinction between organs and tissues. Starting from this point, he discovered remarkable analogies of the tissues, between hairs, feathers and the prickles of the hedgehog; and of the organs, between the arms of a man and the wings of a bird, and between the hands of a man and the claws of a lobster or the trunk of an elephant.

Regarding the assimilation of nourishment by the body, Aristotle held the opinion that foods are cooked by the stomach and are transformed into phlegm or blood according to their degree of cooking.

Finally, Aristotle opened up several new and fruitful paths in embryology, and his observations on teratological cases have not lost their interest. The disciples and successors of Aristotle, although they extended the field of the discoveries made by their master, added nothing to the principles and methods by which he was guided. However, in the vegetable kingdom Theophrastus distinguished the cotyledons (the food leaves contained in the seed) from the ordinary leaves produced on the stem; and recognized the difference of internal structure between palms and other trees.

[1] Quoted from 14 Gomperz, *Penseurs*, III, p. 168.

THE CHEMICAL AND NATURAL SCIENCES 215

Phanias separated flowerless plants such as ferns, mosses and fungi from flowering plants; and it was only eighteen centuries later that this important distinction was revived.[1]

[1] G. Bonnier, *Le monde végétal*, Flammarion, Paris, 1907, p. 38.

CONCLUSION

THE chosen daughter of Zeus, the goddess of the wisdom which inspired war, science and art, Pallas Athene, above all the divinities, was honoured and reverenced by the Athenians; the temple of the Parthenon on the Acropolis symbolizes, even at the present time, the genius of the Greek nation in all its purity. We recall the beautiful prayer of Renan inspired by the sight of this temple: " O nobility, O beauty simple and true, Goddess whose cult symbolizes reason and wisdom, thou, whose temple is an eternal lesson of justice and sincerity, late I come to the threshold of thy mysteries. To find thee has needed an infinity of searching. The initiation which thou didst confer on the newly-born Athenian by a smile, I have won by dint of reflection, at the cost of long struggles."

This homage rendered to the tutelary goddess of Athens expresses in moving words the reverence and gratitude which are inspired by the tremendous labour of civilization accomplished by Ancient Greece. Merely a few centuries have sufficed her, not only for the realization of an incomparable architecture and statuary, but also for the creation of all the known types of literature, and for the establishment of the lasting foundations of most of the sciences. Apparently it was almost without efforts and without gropings in the dark that these conquests were made, in consequence of, as Renan says, the spontaneous initiation granted by reason to every Greek at his birth. In particular, the question arises, How did Ancient Greece succeed in

CONCLUSION

breaking the mental habits of a thousand years, and in forming a true conception of scientific relationships?

Compared with the empirical and fragmentary knowledge which the peoples of the East had laboriously gathered during long centuries, Greek science constitutes a veritable miracle. Here the human mind for the first time conceived of the possibility of establishing a limited number of principles, and of deducing from these a number of truths which are their rigorous consequence.

Beyond the fugitive data of sensation, the Greeks sought for the relationships, which impress the mind as being founded on fact and reason. They were the first to make known the connection of thought and language, and to notice the difference between reasoning and the facts on which it is based.

This work, begun by Parmenides and the sophists, was carried on by Socrates and Plato, and completed by Aristotle. Parmenides caught a glimpse of a realm of truth unshaken by changing opinions; the sophists laid the foundations of grammar; Socrates established the relationship which exists between the general idea and particular ideas contained in it. Plato distinguished two dialectic processes in the realm of thought, the one which proceeds from hypotheses to consequences, the other which starting from hypotheses goes back to the principles which justify them. Finally, Aristotle, in the imposing edifice of his logic, co-ordinates the results obtained by his predecessors. In no other civilization and amongst no other nation do we find any similar systematic and rational analysis of human thought.

Through this analysis the Greeks were led to visualize in every science a matter and a form. The former varies with the object peculiar to each science; the latter is found in every system of reasoned knowledge.

By the form, a consequence is connected to its law

in a necessary manner in the same way as a particular fact to its cause. The objects of science can be classified, as regards their matter, in two groups, according as they arise directly or not from sensible observation.

When the object is not directly related to sensation, as is the case with mathematical facts, the science can be rigorously constituted, because there are a number of primary conceptions from which consequences can be inferred by means of reasoned deduction. For this it is necessary that these primary conceptions should be as logical and as few as possible. The mind is then master both of the form and of the matter of the science, since the latter contains no element foreign to reason.

The sciences which are based upon sensible observation show, like mathematics, a disagreement between form and matter, between a collection of data and a chain of reasoning based on these data. In this case, however, the matter is composed of the individual elements revealed to us by sensation, which can be classified according to the genus, species, etc., to which they belong. In order to make this classification it is necessary in the first place to have recourse to analogical reasoning founded on observation, but, when the classification has once been effected, a deductive syllogism enables each thing to be assigned its place in the universe.

To the Greeks, there was no radical opposition between the inductive and the deductive syllogism. When, having a knowledge of the science, we reason by deduction, we are reproducing the order of nature which creates individuals as a function of the genus and species to which they belong. On the other hand, in order to acquire a knowledge of the science we must start from individual observations and have recourse to inductive syllogism. "Men, horses and mules are long-lived. Now men, horses and mules are animals which have no

gall. Therefore animals without gall are long-lived."
The opposition between induction and deduction, which
has been pointed out in modern times, is not, according
to Aristotelianism, founded upon nature. The unity
of the two perspectives, which, from the standpoint of
critical reflection, appear incompatible, is, according to
Aristotle, ensured by the inversion of the order of progressively acquired knowledge and the order of nature,
" between order as it appears to us and order in itself."
According to a remarkable saying of the Nicomachean
Ethics (1112 b 23), " Τὸ ἔσχατον ἐν τῇ ἀναλύσει, πρῶτον
ἐν τῇ γενέσει." [1] The aim of the sciences which are
based on sensible observation is thus to discover the
classification and natural hierarchies of phenomena in
relation to one another. Their main work is to group
extensively and comprehensively the conceptions to
which these phenomena correspond. The physical
causality which justifies this grouping is imbued with
finality and cannot admit absolute quantitative
relations, except in rare cases.

For the Greeks there existed a cleft between the
mathematical sciences and the physical or natural
sciences, and in their opinion this cleft could never be
closed up. The reason appears to be as follows.

The sciences whose data are exclusively provided by
sensation are concerned with bodies which, with the
exception of the heavenly bodies, are subject to birth,
death, and compulsory motion. These bodies, besides,
obey a cause which displays its effects in time by virtue
of the finality inherent in nature. As individuals they
never realize, except imperfectly, the form towards
which they aspire. Consequently, between form and
matter, there cannot exist an adequate relation,
mathematically measurable, and from a logical point
of view there are always obscurities. Doubtless

[1] L. Brunschvicg, *Expérience humaine et causalité*, p. 157.

nature tends to be penetrated by rationality, but this penetration is never complete because of the resistance which matter offers to form, and this is why individual beings are always imperfect examples of form. The numerical and spatial relations as conceived by arithmetic and geometry have a totally different character, for these relations are eternal, independent of time, of physical place and of circumstances. If, as Aristotle thought, mathematical entities have been gradually disentangled by abstraction from the sensible world, having once been obtained by this process, they appear in a perfect and immutable form. This being so, individual mathematical entities are an exact reproduction of the genus and species to which they belong. Every isosceles triangle, whether small or large, possesses completely and perfectly all the properties of the isosceles triangle, in this sense, that, having two sides equal, it necessarily has two angles equal. Mathematical entities attain their perfect form without any progression in time. The abstract relation of which they are constituted is eternal, or rather it is a relation, irrespective of time, between laws and consequences, in which the efficient cause and the final cause are merged by an indivisible action of the mind.

This fact determines the nature of mathematical conceptions and demonstrations within the following limits :

The primary propositions (axioms, definitions, postulates) must avoid making any appeal to obscure ideas of the sensible intuition such as indefinite dichotomous divisibility and the relation of motion to space.

On the other hand, in geometrical demonstration it is most necessary to use static methods, and to consider as foreign to science the constructions which result from the meeting of two lines in motion.

In the same way, in dealing with integration, the passage to the limit cannot be directly effected. It can

CONCLUSION

only be demonstrated that a curvilinear area is contained between two rectilinear areas whose surfaces differ by a quantity as small as desired. A circle, for example, is contained between the increasing surface of an inscribed polygon and the decreasing surface of a circumscribed polygon.

Because of their distinctive characteristics, it was the mathematical sciences alone which could realize the Greek ideal of axiomatic science, namely a number of logical principles whose rigorous consequences are ensured by reasoned deduction.

The physical and astronomical sciences, in as far as they have attempted to realize this ideal, have been obliged to limit the field of their investigations.

Astronomy, for instance, extricated itself from meteorology, with which it was at first mingled, and attempted, with the Pythagoreans, to unite physics and mathematics. This effort having but imperfectly succeeded, there arose a division between the mechanics of the eternal celestial bodies and that of the terrestrial bodies subject to birth and death. Astronomy then attributed to the celestial bodies a circular motion, and limited its ambition to a geometrical representation of their movement in the heavens. It mattered little whether this representation was physically realizable; it was sufficient that it accounted for the appearances of the celestial phenomena. This being so, the theory of axioms is satisfied, because the circular movement is the only regular and periodic movement which can be logically conceived for a body in space. In fact, if this body did not move circularly, either it would set off at a tangent and go away into infinity, which is impossible in a finite universe; or it would fall to the centre of the universe and everything would be motionless, which is contrary to appearances.

Similar observations apply to mechanics. Being desirous of constituting this science according to an axiomatic type similar to that which characterizes the *Elements* of Euclid, Archimedes confined his studies to statics. In doing this, he thought to find in a purely logical principle—the principle of symmetry—a sufficient foundation for the law of the lever and that of the equilibrium of bodies. If he did not attempt to found dynamics, it was probably for fear of being obliged to have recourse to an obscure sensible intuition. The study of a body in motion implies notions of continuity and indefinite divisibility in time and space, notions which are always in some degree irreconcilable with logic.

Aristotle was more venturesome; but his dynamic theses are rendered obscure by a notion of force which is borrowed from biological conceptions.

Greek science directed along these lines was bound to come to a standstill.

In the first place, the field assigned to mathematics is too restrained and too arbitrary, since the curves called mechanical are excluded from it. Then, within these limits, the demonstrations become more and more complicated from fear of making a direct appeal to infinity. Doubtless the use of infinity offers advantages which are inappreciable from the standpoint of demonstrative rigour, but it is difficult and inconvenient to manipulate, and it lacks generality and necessitates, in its progressive application, more and more complicated geometrical constructions.

This mistrust of infinity, already so great as concerns integration, appears again and in a more marked degree in questions relating to geometrical space. The Greeks refused to think of this as infinite. Consequently they never imagined as possible the geometrical existence of points and straight lines removed to infinity. We know how much these ideas have vivified modern geometry; they have rendered possible generalizations and simplifications of which the Ancients had no conception.

In a quite different direction the physical and natural sciences were likewise arrested in their development. For the conception of finality, upon which they were based by Aristotle, clashes with a difficulty which is clearly emphasized by M. Brunschvicg. The Aristotelian formula leaves the mind undecided between two contrary tendencies : immanence and transcendence. " On the one hand, beings develop by realizing the proper form inherent to them, which is themselves in what is intimate and specific in their reality. On the other hand, this realization implies nevertheless in each being an aspiration to pass beyond its actual state, which cannot be wholly explained except by an attraction towards a higher and in some measure exterior end. The world of spontaneous living beings forms a hierachy turned towards God and of which God Himself, although He does not turn towards the world, is the origin, the prime mover. The doctrine of causation, as it was elaborated by the Aristotelians, oscillates between two tendencies which, if singly developed, would lead to two antagonistic visions of God and the universe." [1]

The Greek conception of the science of axioms is certainly very remarkable, for it accustoms the mind to be very exacting as regards proofs and demonstrations. It evidences, however, an exaggerated prudence and timidity. It not only hampered the progress of mathematics, but it showed itself to be impracticable in the domain of physical science, for the foundations which it specifies for scientific research in this domain are too narrow to support the ideas deduced from experience, such as those of motion, continuity and indefinite divisibility.

Now these notions inevitably appear when one comes

[1] *Expérience humaine et causalité physique*, p. 158, Alcan, 1922.

to closer grips with truth, hence an important problem presents itself : How did the savants of the Renaissance succeed in filling up the gap which, to the Greek mind, existed between physics and mathematics ? How did they succeed in reconciling the requirements of the Greek theory of axioms with the no less irresistible data of experience.

This question may be answered in a few words as follows. As we have seen, Greek science had two requirements :

1. A rigorous chain of propositions ;
2. A collection of ideas which serves as the basis of this chain of reasoning and whose logical truth is convincing to the mind.

The scientists of the Renaissance maintained the first of these two requirements in its integrity, but they partially modified the second.

In every science the connection between propositions must be rigorous, there can be no dispute on this point.

However, the primary notions (axioms, definitions) which form the basis of the reasoned deduction are not necessarily logically clear ; a constant verification by experience is sufficient to make them valid. We do not know, for instance, what motion is in itself, but if we can decompose it into certain measurable elements (time, space), and if this decomposition is useful and accounts for observed facts, we can include it in our primary notions.

By proceeding in this fashion the scientists of the Renaissance succeeded in constituting a science which was both rational and experimental. The aim which they pursued more or less consciously was to make the mathematical conceptions less rigid so as to adapt them to the interpretation of mechanical and physical facts ; and to create a type of law which, whilst allowing of rigorous deductions, expresses the real con-

CONCLUSION

nections of phenomena. The task was immense, and in order to accomplish it successfully it was necessary to surmount difficulties which appeared insoluble.

These difficulties having been overcome, it might have been believed that the way was definitely open and that it was only necessary to advance along it without fear of meeting with insurmountable obstacles. As a matter of fact, until the beginning of the twentieth century, the conception of scientific law, formed by the scientists of the Renaissance, was not seriously shaken. According to this conception, there exist, at the base of all science, rational and experimental laws which having once been discovered are eternally true and incapable of modification. Hence, it is only through the more and more extended application of these laws that science in all its branches will make progress.

We know how the theory of relativity enunciated by Einstein and upheld by Langevin has shaken this conception and put in check certain postulates of the Newtonian kinematics. It is a curious fact that the partial abandonment of the conceptions formulated in the sixteenth and seventeenth centuries marks at the same time a return to many of the opinions held by Greek science in antiquity; this return is all the more significant because it was unpremeditated. It is beyond question that analogies, both as regards hypotheses and methods, can be found between the physics of relativity and the cosmology of the ancient Greeks.

The first philosophers of Ionia, for instance, did not distinguish between an empty space which would be self-existent, and a fluid substance (air, water or fire) which would accidentally fill it. In their eyes there was no separation between the physical properties of space and space itself. In the physics of relativity the same thing occurs in a form, needless to say, infinitely more complex and more justified. There are gravitational and electromagnetic properties which confer on space, in every region, its geometrical qualities (curvature, possible kinds of triangles, etc.).

This being so, there cannot be a universal system of reference, granted once for all, to which the study of a group of localized phenomena, in any part whatever of the universe, can be related. The system of reference must in every case be intrinsic to this group of phenomena, which are then studied by methods necessitating the use of tensorial calculus and absolute differential calculus. As G. Juvet points out, " the characteristic feature of these methods arises from the fact that they enable a geometrical entity to be studied from a purely intrinsic point of view. The Greeks never studied their geometry in any other way, when they were searching for the properties of a figure, they always examined the figure itself, considered by itself and taken independently of any system of reference." [1] It is evident that in Greek geometry, as in the algorithm of relativity, the relations of a figure are sufficient in themselves, and although they may be studied by means of a method and by universal formulæ, it is not necessary, as in Cartesian geometry, to relate them to an exterior system of co-ordinates.

We know, besides, that the universe of the physics of relativity, while lending itself to questions of infinity, remains finite in its dimensions by virtue of its curvature. Now, as we have seen, the hypothesis of finiteness is characteristic of Greek astronomy. As we have pointed out, Empedocles expressed an idea concerning the universe, considered as finite, which recalls that of phantom stars ; he declared in fact that the sun has no real existence, that it is formed by a simple concentration of luminous rays which are

[1] G. Juvet, *Introduction au calcul tensoriel*, A. Blanchard, Paris, 1922.

reflected on the earth and then stopped by the celestial vault.

Another no less interesting analogy to be noted is the following: The so-called theorem of Pythagoras is at the base of the earliest speculations of Greek geometry; it was this which gave rise to the problem of incommensurables and indirectly to the dialectic of Zeno. Now this dialectic is chiefly concerned with the following problem; space, according to the Greeks, is an objective reality which is postulated as motionless. How then is it possible to conceive the relation between a moving object such as an arrow and motionless space?

The difficulty which gave birth to the physics of relativity, and which the Michelson-Morley experiment has brought fully into light, is quite analogous. A source of light, according as it is motionless or moving, ought to behave differently in relation to the ether supposed to be motionless. But as a matter of fact this is not so. How is this to be explained? Here comes in the conception of a spatial-temporal interval and the quadratic expression

$$ds^2 = dx_1^2 + dx_2^2 + dx_3^2 + dx_4^2,$$

which is only a generalized form of the theorem of Pythagoras.

Without investigating the metaphysical range and practical use of this fusion of space and time, the important fact remains that the physics of relativity, considered theoretically, is a remarkable attempt to constitute a theory of axioms comparable to that of Euclid. Only this attempt does not aim at establishing the domain of a mathematics which is separated from reality; it tends to unite in one whole the geometrical, mechanical and physical properties of the universe. Evidently, as Winter points out, such a science of axioms cannot pretend to create logically and *a priori* the real world apart from experience; it can only analyse, that is to say elaborate, the group of axioms necessary and sufficient to explain real phenomena.[1] The axiomatic analysis as thus understood seeks to substitute clear and distinct ideas for intuitive, experimental and often confused notions. It is thereby carrying on not only the work of Descartes but also that of Greek science.

Hence we are forced to the following conclusion: the physics of relativity in returning to the immediate data of sensible experience seeks to reduce them to axioms, and for this reason it comes into line with the realistic and logical tendencies of the Greek thinkers of antiquity.

[1] *Revue de Métaphysique et de Morale*, "The Theorem of Pythagoras," p. 23, year 1923.

BIBLIOGRAPHY

The list of the principal publications relating to Egypt and Chaldea may be found in the following works : G. JÉQUIER, *Histoire de la civilisation égyptienne.* 2nd edit. Payot, Paris 1923.—L. DELAPORTE, *La Mésopotamie ; les civilisations babyloniennes et assyriennes.* Bibliographie de synthèse historique. Renaissance du Livre. Paris 1923.

A biography of both Greek and Roman science is given by L. LAURAND, *Manuel des études grecques et latines.* Appendix I, *Les sciences dans l'Antiquité.* Picard, Paris 1923. For a more complete systematic account see the chapter on the exact sciences and medicine by M. HEIBERG in the work : A. Gercke and E. Norden, *Einleitung in die Altertumswissenschaft.* 3rd edit. Teubner, Leipzig 1922.

It must be remembered that the history of sciences in the past presents peculiar difficulties. Considering that their character was purely objective, tradition often omits to mention the names of those who made or perfected certain discoveries. As the sciences progressed, the works of an earlier period, except those which had become classics, were constantly absorbed by works which a later period considered superior. Thus the earlier works disappeared and their titles scarcely have been preserved. Under these conditions, it is very difficult to accurately determine the date and origin of the information handed down to us by the scientific literature of antiquity. We therefore limit ourselves to certain indications.

History of the Sources.—H. DIELS, *Doxographi graeci.* Reimer. Berlin 1879.—J.-L. HEIBERG, work cited, *Einleitung in die Altertumswissenschaft.*—J.-L. HEIBERG, *Les sciences grecques et leur transmission.* Scientia, vol. XXXI, p. 1 and p. 97.—F. UEBERWEG, *Grundriss der Geschichte der Philosophie,* 11th edition, p. 16 *et seq.* Vol.

I. Mittler und Sohn, Berlin 1920.—L. ROBIN, *La pensée grecque et les origines de l'esprit scientifique*. Bibl. de synthèse historique. Renaissance du Livre, Paris 1923. Chapter II, p. 8 *et seq*.

General Works.—DAREMBERG and SAGLÍO, *Dictionnaire des Antiquités grecques et romaines*. Hachette, Paris 1873–1919.—PAULY-WISSOWA, *Real-Encyclopaedie der klassischen Altertumswissenschaften*. 2nd edit. 1884 in course of publication, 13 vol.—J.-L. HEIBERG, *Naturwissenschaft und Mathematik im klassischen Altertum*. Teubner, Leipzig 1912.—G. LORIA, *Le Scienze esatte nell' Antica Grecia*, 2nd edit. Hoepli, Milan 1914.—TH. GOMPERZ, *Les penseurs de la Grèce*. Translation Aug. Reymond. Vol. I–III. Payot, Lausanne 1904–1910.—P. TANNERY, *Mémoires scientifiques : Sciences exactes dans l'antiquité*. 3 vol. Privat, Toulouse ; Gauthier-Villars, Paris 1912.— G. MILHAUD, *Etudes sur la pensée scientifique chez les Grecs et chez les modernes*. Alcan, Paris 1906.—G. MILHAUD, *Nouvelles études sur l'histoire de la pensée scientifique*. Alcan, Paris 1911.

Pre-Socratic Period.—H. DIELS, *Die Fragmente der Vorsokratiker*. Vol. I and II, 2nd edit. Weidmann, Berlin 1906–1907.—P. TANNERY, *Pour l'histoire de la Science hellène*. Alcan, Paris 1887.—G. MILHAUD, *Les philosophes géomètres de la Grèce. Platon et ses prédécesseurs*. Alcan, Paris 1900.—G. MILHAUD, *Leçons sur les origines de la science grecque*. Alcan, Paris 1893.—J. BURNET, *L'aurore de la philosophie grecque*. Translation Aug. Reymond. Payot, Paris 1919.—A. RIVAUD, *Le problème du devenir et la notion de matière dans la philosophie grecque depuis les origines jusqu'à Théophraste*. Alcan, Paris 1906.

MATHEMATICAL SCIENCES : GEOMETRY AND ARITHMETIC

(*a*) **Modern Works.**—M. CANTOR, *Vorlesungen uber Geschichte der Mathematik*. Vol. I, 3rd edition. Teubner, Leipzig 1894.—H.-G. ZEUTHEN, *Histoire des mathématiques dans l'antiquité et le moyen âge*. Translation Jean Mascart. Gauthier-Villars, Paris 1902.—H.-G. ZEUTHEN, *Die mathe-*

matischen Wissenschaften (Kultur de Gegenwart III, Abt. I, 1). Teubner, Leipzig 1912.—T.-L. HEATH, *A history of Greek mathematics.* 2 vols. Clarendon Press, Oxford 1921.—W.-W. ROUSE BALL, *Histoire des mathématiques.* Translation L. Freund, 2 vol. Hermann, Paris 1906.— M. MARIE, *Histoire des sciences mathématiques et physiques.* 12 vol. Gauthier-Villars, Paris 1883-1888.—J.-F. MONTUCLA, *Histoire des mathématiques.* 4 vol. Agasse, Paris 1799-1802.—P. TANNERY, *La géométrie grecque.* Gauthier-Villars, Paris 1887.—P. BOUTROUX, *Les principes de l'analyse mathématique, exposé historique et critique.* 2 vol. Hermann, Paris 1914.—P. BOUTROUX, *L'idéal scientifique des mathématiciens.* Alcan, Paris 1920.—M. CHASLES, *Aperçu historique sur l'origine et le développement des méthodes en géométrie.* 2nd edit. Gauthier-Villars, Paris 1875.—L. BRUNSCHVICG, *Les étapes de la philosophie mathématique.* Alcan, Paris 1912.

(*b*) **Ancient Authors.**—*Geometry.*—The history of geometry has been transmitted by EUDEMUS (L. Spengel, *Eudemii Rhodii Peripatetici fragmenta,* Berlin 1866.— F.-S.-A. Mullach, *Fragmenta philosophorum graecorum,* vol. II, p. 266. Didot, Paris 1883) and by PROCLUS (G. Friedlein, *In primum Euclidis Elementorum librum Procli commentarii.* Teubner, Leipzig 1873). Much information is also to be found in the works of PAPPUS.

EUCLID : Greek and Latin text. 8 vol. [*Elementa* edited by J.-H. Heiberg, 5 vol. 1883-88. *Data* by H. Menge, 1896. *Optica et catoptrica* by J.-L. Heiberg, 1895. *Phaenomena et scripta musica,* by H. Menge ; *Fragmenta,* by J.-H. HEIBERG, 1916. Teubner, Leipzig.]

ARCHIMEDES : Greek and Latin text, 2nd edition, by J.-L. Heiberg, 1910. 3 vol. Teubner, Leipzig.—*Les Œuvres complètes d'Archimède,* translated into French by Paul ver Eecke. Desclée de Brouwer et Cie, Paris, Bruxelles 1921.—*The works of Archimedes,* edited in modern notation with introductory Chapters by T. L. Heath. University Press, Cambridge 1897.

APOLLONIUS : Greek and Latin text. *Libri I–IV cum commentariis antiquis,* 2 vol., by J.-L. Heiberg, 1891-93. Teubner, Leipzig.

PAPPUS ALEXANDRINUS, *Collectio*, Greek and Latin text, edited by F. Hultsch 1876-78. 3 vol. Weidmann, Berlin.

Arithmetic.—On the history of this teaching we possess only a little information contained in the books VII-IX of the *Elements* of Euclid.—NICOMACHUS, *Introductio arithmetica*, Greek text, by R. Hoche, 1866.—THEON OF SMYRNA, *Expositio rerum mathematicarum ad legendum Platonem utilium*, Greek text, by E. Hiller, 1878. Teubner, Leipzig.—IAMBLICHUS, *In Nicomachi arithm. introd.*, Greek text, by E. Pistelli, 1894. Teubner, Leipzig.—IAMBLICHUS, *De communi mathematica scientia*, Greek text, by N. Festa, 1891. Teubner, Leipzig.—DIOPHANTUS, Greek and Latin text, edited by P. Tannery, 2 vol. 1893-95. Teubner, Leipzig.—BOETIUS, *Institutio arithmetica et institutio musica*, edited by G. Friedlein, 1867. Teubner, Leipzig.

Physics and Mechanics.—Among modern authors, consult : F. ROSENBERGER, *Die Geschichte der Physik in Grundzügen*, vol. I. Braunschweig 1882.—A. HELLER, *Geschichte der Physik von Aristoteles bis auf die neueste Zeit*, vol. I. Stuttgart 1882.—H. DIELS, *Antike Technik*, 2nd edit. Teubner, Leipzig 1920.—A. DE ROCHAS, *La science des philosophes et l'art des thaumaturges dans l'antiquité*. Dorbon, Paris.—L. ROBIN, *Etude sur la signification et la place de la physique dans la philosophie de Platon*. Alcan, Paris 1919.—A. MANSION, *Introduction à la physique aristotélicienne*. Louvain 1913.—G. RODIER, *La physique de Straton de Lampsaque*. Alcan, Paris 1890.—H. DIELS, *Ueber das physikalische System des Straton*. Sitzungsberichte Berlin. Akad. 1893. p. 101.—E. JOUGUET, *Lectures de mécanique*, 2 vol. Gauthier-Villars, Paris 1908-9.—P. DUHEM, *Les origines de la statique*, 2 vol. Hermann, Paris 1905-6.—H. CARTERON, *La notion de force dans le système d'Aristote*. Vrin, Paris 1924. This learned work, issued whilst the present book was in the press, emphasizes the metaphysical and scientifically inutilizable nature of the mechanical propositions formulated by Aristotle.

HERO OF ALEXANDRIA is our principal source for theoretical and practical mechanics. 5 vols. [*Pneumatica et*

automata, Greek-German text, by W. SCHMIDT, 1899.—
Mechanica, Arabic-German, *and Catoptrica,* Greek-German, by L. NIX and W. SCHMIDT, 1900.—*Rationes dimetiendi et commentatio dioptrica,* Greek-German, by H. SCHONE, 1903. —*Definitiones,* Greek-German, by J.-L. HEIBERG, 1912.— *Stereometrica et de mensuris,* Greek-German, by J.-L. HEIBERG, 1914.] Teubner, Leipzig.—ARCHIMEDES : His writings on mechanics are found in vol. II of his works.

Astronomy.—(*a*) *Modern Authors.*—J.-B. DELAMBRE *Histoire de l'astronomie ancienne,* 1817.—P. TANNERY, *Recherches sur l'histoire de l'astronomie ancienne.* Paris 1893.—E. DOUBLET, *Histoire de l'astronomie.* Doin, Paris 1922. (In this work will be found an excellent appreciation of the lives and the works of the historians of astronomy : WEIDLER, BAILLY, LALANDE, DELAMBRE, MONTUCLA, BOSSUT, BIOT, P. TANNERY, P. DUHEM.)—G.-V. SCHIAPARELLI, *I precursori di Copernico nell' antichita.* Hoepli, Milan 1873. *Le sfere omocentriche di Eudosso, di Calippo e di Aristotele,* 1875.—P. DUHEM, *Le système du monde. Histoire des doctrines cosmologiques de Platon à Copernic,* 5 vol. Hermann, Paris 1913-1917.—J. SAGERET, *Le système du monde. Des Chaldéens à Newton.* Alcan, Paris 1913.—G. BIGOURDAN, *l'Astronomie.* Flammarion, Paris 1917.—J. HARTMANN, *Astronomie (Die Kultur der Gegenwart, III,* 3. 3). Teubner, Leipzig 1921.

(*b*) *Ancient Authors.*—The information concerning the history of astronomy is found particularly in EUDEMUS (L. Spengel edition, Berlin 1866.—G.-A. MULLACH, *Fragmenta phil. graecorum,* vol. III, p. 276) and in ARISTOTLE, *De caelo,* II, 12, as also in SIMPLICIUS, the *Simplicius in Aristotelis.*

AUTOLYCUS, *De sphaera quae movetur ; de ortibus et occasibus,* edited by F. Hultsch 1885. Teubner, Leipzig.— HIPPARCHUS, *In Arati et Eudoxi phaenomena commentarii,* Greek-German text, edited by C. Manitius 1894. Teubner, Leipzig.—CLEOMEDES, *De motu circulari corporum caelestium,* Greek-Latin text, edited by H. Ziegler 1891. Teubner, Leipzig.—GEMINUS, *Elementa astronomiae,* Greek-German text, edited by C. Manitius 1898. Teubner, Leipzig.—PTOLEMY, *Syntaxis mathematica,* 2 vol., Greek text, by J.-L. Heiberg 1898–1903. *Opera astronomica minora*, Greek text and Greek-German text for certain parts, by J.-L. Heiberg, 1908. Teubner, Leipzig.

PROCLUS, *Hypotyposis astronomicarum positionum*, Greek-German text, edited by C. Manitius 1909. Teubner, Leipzig.

On *Astrology* consult A. BOUCHÉ-LECLERQ, *l'Astrologie grecque*. Paris 1899.—FIRMICUS MATERNUS, *Matheseos*, Latin text, edited by W. Kroll, F. Skutsch and K. Ziegler 1897. Leipzig, Teubner.—MANILUS, *Astronomica*, Latin text, by T. van Wageningen 1915. Leipzig, Teubner.— *Catalogus codicum astrologorum graecorum*, by F. BOLL, F. CUMONT, etc., in course of publication, 8 vol. issued Brussels 1898–1911.

Geography.—ERATOSTHENES, *Geograph. fragm.* edited by H. Berger 1880.—HIPPARCHUS, *Geograph. fragm.* edited by the same 1869. Teubner, Leipzig.—STRABO, *Geographica*, text edited by A. Meinecke 1866. Teubner, Leipzig.— Consult H. BERGER, *Die Geschichte der wissenschaftlichen Erdkunde der Griechen*, 2nd edit. 1903. Leipzig.

Chemistry.—For the texts see *Collection des anciens Alchimistes grecs*, 3 vol., by M. BERTHELOT and C. RUELLE. Steinheil, Paris 1888.—M. BERTHELOT, *Introduction à l'étude de la chimie des anciens et du moyen âge*. Steinheil, Paris 1889.—M. DELACRE, *Histoire de la chimie*. Gauthier-Villars, Paris 1920.—H. DIELS, *Antike Technik*, 2nd edit. Teubner, Leipzig 1920.

Natural Sciences.—General study, V. HEHN, *Kulturpflanzen und Haustiere in ihrem Uebergang aus Asien nach Griechenland und Italien*, 8th edit. by O. Schrader Borntraeger. Berlin 1911.

Zoology.—ARISTOTLE, *Opera omnia*, Academia regia borussica. Vol. 1 and 2, Greek text; vol. 3, Latin translation; vol. 4, *Scholies*; vol. 5, *Fragments scholies, Index*. Reimer, Berlin, 1831–1870.—ÆLIANUS, *De natura animalium*, 2 vol. Text edited by R. Hercher 1864. Teubner, Leipzig.

To consult : V. CARUS, *Histoire de la zoologie*, French translation. Baillière, Paris 1880.—E. PERRIER, *La philosophie zoologique avant Darwin*, 3rd edit. Alcan, Paris 1896.

BIBLIOGRAPHY

Botany.—THEOPHRASTUS, *De plantarum historia; de plantarum causa*, by F. Wimmer, 3 vol. Teubner, Leipzig 1854.—DIOSCORIDES, *De materia medica*, by M. Wellmann. Berlin 1906, 2 volumes issued.

Medicine.—An immediate disciple of Aristotle, MENON, wrote a history of medicine the fragments of which have been edited by H. Diels, *Sup. Aristo.* vol. III, Berlin 1893.— In modern times, consult: L. MEUNIER, *Histoire de la médecine*. Paris, Baillière 1911.—TH. PUSCHMANN, M. NEUBURGER, J. PAGEL: *Handbuch der Geschichte der Medizin*. Fischer, Jena 1902, vol. I (Greek medicine by R. Fuchs; Roman medicine by Iw. Bloch).

The texts of the authors are not yet established with all the desirable philological care (see on these gaps J.-L. HEIBERG, *op. cit.*: *Einleitung in die Altertumswiss.*, p. 413).—A revision of the works of HIPPOCRATES has been undertaken by H. Kuehlewein and I. Ilberg. Teubner, Leipzig. 2 vol. issued.—Meanwhile, the edition of Littré with French translation remains the best source. 10 vol. Baillière, Paris 1839–61.

GALEN, *Scripta minora*. Greek text edited by J. Marquardt, I. Müller, G. Helmreich. Teubner, Leipzig 1884–1893.—C. CELSUS, *De re Medica*, edited by C. Daremberg. Teubner, Leipzig 1891.—SORANUS, *Gynaecia*. Ancient Latin translation and Greek text, by W. Rose. Teubner, Leipzig 1882.—RUFUS of Ephesus, Greek and French text, edited by Ch. Daremberg and E. Ruelle. Baillière, Paris 1854–56.—A large edition of Greek and Roman physicians is in course of publication by Teubner, Leipzig. *Corpus medicorum graecorum; corpus medicorum latinorum.*

LIST OF THE PRINCIPAL WORKS MENTIONED

1. BERTHELOT (M.).—*Introduction à la chimie des Anciens et du Moyen Age.* Steinheil, Paris 1889.
2. BIGOURDAN (G.).—*L'Astronomie. Evolution des idées et des méthodes.* Flammarion, Paris 1911.
3. BOUTROUX (P.).—*Les principes de l'analyse mathématique, exposé historique et critique.* 2 vol. Hermann, Paris 1914.
4. BOUTROUX (P.).—*L'idéal scientifique des mathématiciens.* Alcan, Paris 1920.
5. BOUTROUX (P.).—*Les mathématiques.* Albin Michel, Paris 1922.
6. BOYER (J.).—*Histoire des mathématiques.* Carré et Naud, Paris 1900.
7. BRUNSCHVICG (L.).—*Les étapes de la philosophie mathématique.* Alcan, Paris 1922.
8. BURNET (J.).—*L'aurore de la philosophie grecque.* Translation Aug. Reymond. Payot, Paris 1919.
9. CANTOR (M.).—*Vorlesungen über Geschichte der Mathematik.* Vol. I, 2nd edit. Teubner, Leipzig 1894.
10. DIELS (H.).—*Antike Technik.* Teubner, Leipzig 1914.
11. DUHEM (P.).—*Les origines de la statique.* 2 vol. Hermann, Paris 1905-1906.
12. DUHEM (P.).—Σωζειν τα φαινομεγα. *Essai sur la notion de théorie physique de Platon à Galilée.* Hermann, Paris 1908.
13. DUHEM (P.).—*Le système du monde. Histoire des doctrines cosmologiques de Platon à Copernic.* 5 vol. Hermann, Paris 1913-1917.
14. GOMPERZ (TH.).—*Les penseurs de la Grèce.* Translation Aug. Reymond. 3 vol. Payot, Lausanne 1904-1910.

15. HEIBERG (J.-L.).—*Naturwissenschaften und Mathematik im klassischen Alterthum.* Teubner, Leipzig 1912.
16. JOUGUET (E.).—*Lectures de mécanique.* 2 vol. Gauthier-Villars, Paris 1908–1909.
17. LORIA (G.).—*Le Scienze esatte nell' Antica Grecia.* 2nd edit. Hoepli, Milan 1914.
18. MASPERO (G.).—*Histoire ancienne des peuples de l'Orient.* 3rd edit. Hachette, Paris.
19. MASPERO (G.).—*Lectures historiques.* 4th edit. Hachette, Paris 1905.
20. MILHAUD (G.).—*Les philosophes géomètres de la Grèce. Platon et ses prédécesseurs.* Alcan, Paris 1900.
21. MILHAUD (G.).—*Études sur la pensée scientifique chez les Grecs et chez les modernes.* Alcan, Paris 1906.
22. MILHAUD (G.).—*Nouvelles études sur l'histoire de la pensée scientifique.* Alcan, Paris 1911.
22a. ROBIN (L.).—*La pensée grecque et les origines de l'esprit scientifique.* Renaissance du Livre, Paris 1923.
23. ROUSE BALL (W.).—*Histoire des mathématiques.* Translated by L. Freund. 2 vol. Hermann, Paris 1906.
24. SAGERET (J.).—*Le système du monde.* Alcan, Paris 1913.
25. TANNERY (P.).—*Pour l'histoire de la science hellène.* Alcan, Paris 1887.
26. TANNERY (P.).—*La géométrie grecque, comment son histoire nous est parvenue et ce que nous en savons.* Gauthiers-Villars, Paris 1887.
27. TANNERY (P.).—*Recherches sur l'histoire de l'astronomie ancienne.* Gauthier-Villars, Paris 1893.
28. TANNERY (P.).—*Mémoires scientifiques.* 3 vol. Toulouse et Gauthier-Villars, Paris 1912.
29. ZEUTHEN (H.-G.).—*Histoire des mathématiques dans l'antiquité et le moyen âge.* Translation J. Mascart. Gauthier-Villars 1902.
30. ZEUTHEN (H.-G.).—*Die mathematischen Wissenschaften* (*Kultur der Gegenwart*, III, Abt. I, 1). Teubner, Leipzig 1912.

PRINCIPAL WORKS MENTIONED

Texts of the Greek and Latin authors: *Bibliotheca Teubneriana*. Teubner, Leipzig.—Special editions: H. DIELS, *Doxographi graeci*. Reimer, Berlin 1879.—H. DIELS, *Die Fragmente der Vorsokratiker*, 2nd edition. Weidmann, Berlin 1906.—ARISTOTELIS OPERA, 5 vol., Berlin academical edition. Reimer, Berlin 1831.—PAPPI ALEXANDRINI, *Collectionis quae supersunt*. F. Hultsch edition, 3 vol. Weidmann, Berlin 1876.

INDEX

Ælianus, 234
Aetius, 27, 28, 165
Agrippa, 92, 95
Ahmes, 2, 4, 5
Al-Bitrogi, 175
Alcmæon, 36, 49, 52, 53, 214
Alexander the Great, 19, 61, 64, 65, 81
Ameinias, 37
Anaxagoras, 43, 46, 163, 164
Anaximander, 22, 25, 26, 27, 28, 38, 163, 164
Anaximenes, 22, 28, 29, 44, 163, 164
Antiphon, 56, 131
Antonines, 95
Apollonius of Citium, 90
Apollonius of Perga, 66, 69, 76, 77, 78, 98, 99, 113, 144, 147, 159, 174, 231
Aratus, 60, 85
Archigenes, 101
Archimedes, 60, 66, 70, 71, 72, 75, 76, 77, 78, 80, 81, 83, 85, 98, 99, 113, 126, 134, 135, 136, 180, 192–9, 200, 222, 231, 233
Archippus, 34
Archytas, 34, 59, 61, 62, 72, 179
Aretæus, 101
Aristarchus of Samos, 75, 84, 88, 173, 176
Aristo, 57, 69
Aristobulus, 64
Aristophanes of Byzantium, 89
Aristophanes, 56, 182
Aristotle, 44, 45, 60, 61–3, 64, 70, 89, 95, 131, 132, 159, 169, 172, 173, 183–92, 199, 200, 201, 207, 208, 212–14, 217, 219, 220, 222, 223, 232, 233, 234.—*Phys.* 38, 128, 131, 185, 186, 187, 190.—*De coelo*, 132, 189.—*De gener. anim.* 212, 214.—*Meteor,* 62.—*Quest. mechan.* 192.—*Ethic Nic.* 213
Aristoxenus, 34, 63
Asclepiades, 90, 100
Athenæus, 100
Attalus, 76
Augustus, 87, 94
Aulus Gellius, 180
Aurelianus, Caelius, 94
Autolycus, 64, 233
Averroes, 175

Bailly, 233
Berger, H., 234
Berthelot, M., 204, 205, 234, 237
Berthelot, R., 183
Bigourdan, G., 8, 11, 12, 13, 14, 23, 87, 97, 173, 233, 237
Biot, J. B., 233
Boetius, 93, 97, 232
Bolos, 204
Bonnier, G., 215
Bosse, 159
Bossut, 233
Bouché-Leclerq, A., 234

242 SCIENCE IN GRECO-ROMAN ANTIQUITY

Boutroux, E., 213
Boutroux, P., 113, 116, 117, 121, 123, 141, 144, 149, 153, 156, 157, 231, 237
Boyer, J., 5, 69, 93, 237
Breasted, J. H., 16
Bryson of Heraclea, 56
Bruno, G., 176
Brunschvicg, L., 63, 106, 122, 219, 223, 231, 237
Buffon, 71
Burnet, J., 22, 23, 26, 29, 31, 32, 33, 34, 35, 39, 40, 41, 42, 47, 163, 230, 237

Cadmus, 17
Calippus, 62, 173
Callimachus, 89
Cambyses, 14
Cantor, M., 4, 93, 230, 237
Carpus, 155, 156
Carteron, H., 183, 232
Carus, V., 234
Cato, 90
Cavalieri, 136
Cebes, 34
Celsus, C., 89, 94, 235
Censorinus, 93
Cæsar, 95
Chasles, M., 69, 160, 231
Chemes, 205
Cicero, 31, 64, 71, 88, 92, 167
Cleanthes, 84
Clement of Alexandria, 7
Conon, 71, 77, 85
Constantine, 19
Copernicus, 84, 167, 173, 176
Cratevas, 91
Crœsus, 23, 118
Croiset, A., 101
Ctesibius, 78, 79, 200
Cylon, 34

Darius, 49
Darwin, 214

Delacre, M., 203, 234
Delambre, J. B., 85, 233
Delaporte, L., 229
Demetrius of Phalerus, 66
Democedes, 49
Democritus, 22, 46, 48, 64, 131, 204, 207
Dionysius of Syracuse, 34, 180
Desargues, 159
Descartes, 98, 117, 150, 228
Dicearchus, 64
Diels, H., 232.—*Dox.* 25, 27, 28, 29.—*Vor.* 32, 35, 40, 42, 56, 57, 164, 165.—*Antike.* 79, 179, 180, 204, 205, 237
Diocles, 54, 78, 117
Diocletian, 205
Diodorus of Sicily, 1, 72
Diogenes Laertius, 23, 33, 37, 38
Diophantus, 99, 98, 113, 126, 232
Dioscorides, 102, 235
Dositheus, 71, 85
Doublet, E., 86, 87, 233
Duhem, P., 165, 167, 168, 171, 175, 176, 184, 185, 186, 187, 188, 190, 191, 192, 195, 199, 200, 201, 232, 237

Ecphantus, 166, 173
Einstein, 42, 225
Eisenlohr, A., 2
Eleatics, 22, 36–9, 46, 48, 127
Empedocles, 29, 39–43, 46, 51, 162, 206, 207, 214, 226
Erasistratus, 89, 90
Eratosthenes, 28, 81–3, 87, 234
Euclid, 57, 60, 66–70, 77, 78, 93, 96, 98, 99, 114, 118, 121, 125, 150–8, 182, 193, 222, 231
Eudemus, 57, 63, 76, 231
Eudoxus of Cnidus, 57, 60, 62, 68, 84, 132–4, 171, 172, 174

INDEX

Eupalinus, 179
Euripides, 181
Eutocius, 73, 99

Facundus Novus, 12
Fermat, 136
Flammarion, C., 11

Galen, 54, 101-2, 235
Galileo, 177
Gelo, 70
Geminus, 78, 88, 155, 156, 233
Glotz, G., 17
Gomperz, Th., 36, 39, 49, 50, 210, 211, 212, 213, 214, 230, 237
Guldinus, 98, 200

Hartmann, J., 233
Heath, T. L., 70, 231
Hegel, 21, 39
Hehn, V., 234
Heiberg, J. L., 49, 51, 53, 62, 64, 65, 70, 77, 78, 79, 81, 84, 89, 92, 94, 96, 100, 230, 238
Heller, A., 232
Heraclides of Pontus, 64, 84, 173
Heraclitus, 20, 21, 30-32, 35, 48, 162, 163
Herodotus, 1, 17, 23, 49, 84, 118
Hero of Alexandria, 79-81, 122, 180, 181, 200, 232
Herophilus, 89, 90
Hicetas, 166, 167
Hiero, 70, 72, 75
Hipparchus, 10, 85-7, 97, 174, 175, 233
Hippias of Elis, 55, 117
Hippocrates of Chios, 57-8
Hippocrates of Cos, 49-54, 90, 102, 211, 214, 235
Hippolytus, 28, 29, 164
Houssay, F., 212

Iamblichus, 33, 99
Isidore of Miletus, 80, 99, 232

Jéquier, G., 1, 229
Jordanes, 201, 202
Jouquet, E., 186, 192, 232, 238
Julian the Apostate, 102
Juvet, G., 226

Kepler, 177
Kidinnu, 14

Lalande, 233
Lamarck, 214
Langevin, P., 225
Larguier des Bancels, 33
Laurand, L., 204, 229
Lavoisier, 209
Lecornu, L., 196
Leibnitz, 76, 136
Léon, 60
Lespagnol, G., 82
Leucippus, 22, 43, 46, 131
Lévy-Bruhl, L., 106, 108
Littré, 50, 53, 235
Livy, 70
Lobatschewsky, 154
Loria, G., 18, 22, 69, 79, 182, 230, 238

Mach, E., 193
Manitius, M., 233
Mansion, A., 183
Marcus Aurelius, 102
Marcellus, 70
Marie, M., 231
Martianus Capella, 92
Maternus Firmicus, 93, 205, 234
Maspero, G., 9, 15, 16, 238
Méautis, G., 33
Melissus of Gamos, 39
Menaechmus, 57, 61, 69, 155
Menelaus, 97

Menon, 63, 235
Meton, 57
Meunier, L., 235
Meyerson, E., 112, 119, 183
Michelson, 227
Milhaud, G., 3, 5, 6, 8, 10, 18, 39, 62, 122, 124, 131, 179, 230, 238
Mithridates, 90, 91
Mnesarchus, 33, 35
Montucla, J. F., 231, 233
Morgan, J. de, 203
Morley, 227
Mullach, F. S. A., 231, 233

Nearchus, 64
Neuburger, M., 235
Newton, 83, 136, 170
Nicander, 91
Nicomachus of Gerasa, 93, 97, 126, 232
Nicomachus, 219
Nicomedes, 78, 117
Nigridius Figulus, 93
Nietzsche, F., 39
Nordmann, C., 42

Oribasius, 102
Ostwald, W., 208

Pagel, J., 235
Painlevé, P., 76
Pappus, 56, 76, 98, 149, 150, 200, 232, 239
Parmenides, 37–9, 40, 46, 164, 217
Pascal, 136
Pauly-Wissowa, 230
Pericles, 44
Perrier, E., 234
Phanias, 215
Pheidias, 71
Philolaus, 34, 36, 165, 166
Philo of Byzantium, 79, 80, 200

Philoponus, 57, 55, 201
Physiologus, 103
Piaget, J., 109
Picard, 83
Plato, 34, 46, 55, 57, 59, 62, 68, 88, 96, 152, 168, 169, 181, 182, 207, 213.—*Ion*. 176.—*Laws*, 116.—*Parm*. 37, 38, 39, 128.—*Phaedo*, 34, 44.—*Rep*. 114, 120.—*Theaetetus*, 55.—*Timaeus*, 162, 182
Pliny the Elder, 12, 86, 87, 94
Plutarch, 9, 42, 70, 72
Polemon, 88
Polybius, 70
Polycrates, 33
Pompey, 88
Pomponius Mela, 95
Porphyry, 33, 99
Posidonius, 88, 94
Praxagoras, 54, 89, 90
Praxiades, 25
Proclus, 1, 6, 17, 56, 60, 72, 99, 150, 151, 155, 231, 234
Ptolemy, Claudius, 95–7, 101, 175, 182, 233
Ptolemy I (Soter), 66
Ptolemy II (Philadelphus), 66, 83, 89
Ptolemy III (Euergetes), 85
Ptolemies (Dynasty), 19, 66, 81, 89, 204
Pythagoras, 18, 22, 30, 32–5, 38, 49, 68, 111, 118, 127, 139, 142, 163, 165, 179, 227, 228
Pythagoreans, 21, 41, 48, 57, 88, 118, 121–5, 127, 128, 133, 136, 164, 167, 221

Regiomontanus, 98
Reinach, Th., 76
Renan, E., 216
Reymond, A., 106, 127
Riemann, 154

INDEX

Ritter and Preller, 38
Rivaud, A., 230
Robin, L., 22, 27, 33, 124, 212, 232, 238
Rochas, A. de, 79, 182, 232
Rodier, G., 64, 232
Rouse Ball, W., 2, 58, 67, 76, 83, 125, 127, 231, 238
Rosenberger, F., 232
Ruelle, C., 234
Rufus of Ephesus, 101, 235

Sageret, J., 12, 13, 176, 186, 188, 233, 238
Sallustus, 95
Sardanapalus IV, 5
Schiaparelli, 171, 233
Schmidt, W., 79
Seleucids, 81
Seleucus of Seleucis, 84, 88
Seneca, 94, 182
Serenus, 98
Simmias, 34
Simplicius, 38, 56, 99, 168, 188, 233
Smith, E., 16
Socrates, 34, 217
Soranus, 94, 100, 235
Sorel, G., 118
Speusippus, 155
Stevinus, 202
Stobæus, 35
Stoics, 30
Strabo, 1, 28, 87, 234
Strato of Lampsacus, 64, 66, 84, 90

Tacitus, 95
Tannery, P., 10, 13, 19, 22, 25, 26, 27, 29, 30, 31, 39, 41, 45, 56, 59, 68, 79, 95, 99, 117, 121, 126, 133, 151, 155, 164, 230, 231, 233, 238
Teichmüller, 25, 27
Thales, 22–4, 30, 118, 163, 179
Theaetetus, 57, 68
Themison, 100
Theodorus of Cyrene, 55
Theon of Smyrna, 98, 232
Theon, 99
Theophrastus, 25, 43, 63, 64, 89, 167, 214, 235
Theudius, 60
Thomas Aquinas, 176
Tycho-Brahe, 176

Ueberweg, F., 229

Ver Eecke, P., 70, 72, 231
Vinci, Leonardo da, 201
Vitruvius, 75, 94, 182
Von Lichtenberg, R., 17

Weidler, 233
Winter, M., 227

Xenophanes, 23, 29, 35, 36, 37, 163

Zeller, E., 45
Zeno of Elea, 18, 38, 39, 120, 127–32, 135, 137, 159, 199, 227
Zeuthen, H. G., 2, 4, 5, 6, 67, 129, 157, 238
Zosimus, 103, 205